German Armored R...
1935-1945

Neubaufahrzeug, Luchs, Flammpanzer, Tauchpanzer, Krokodil, Leopard, Löwe, Bär, and many other experimental vehicles and armored projects

by Michael Sowodny

For propaganda purposes, three of the impressive-looking Neubaufahrzeuge, already built in the early thirties but having only meager combat value, were shipped to Oslo during the Norwegian campaign. (BA)

Schiffer Military/Aviation History
Atglen, PA

CREDITS

Federal Archives-Photo Archives, Koblenz (BA)
Federal Archives-Military Archives, Freiburg (BA)
Imperial War Museum, London (IWM)
Royal Armored Corps Tank Museum, Bovington (RAC)
Nowarra Archives
Podzun Archives
H. L. Doyle (drawings)
Heiko Scheibert (drawings)

My special thanks go to the Federal Archives-Military Archives in Freiburg, which freely allowed me to examine many not yet evaluated original documents of the Army Weapons Office and of various firms. Only thus could I assemble enough text and, above all, illustrative material for the book before you. Thanks also to Prof. Dr. Sawodny and my brother Oliver Sawodny for all their editorial assistance to me.

Translated from the German by Ed Force

Printed in China.
ISBN: 0-7643-0396-1

This book was originally published under the title,
Deutsche Panzer-Raritäten 1935-1945: Neubaufahrzeug, Luchs, Flammpanzer, Tauchpanzer, Krokodil, Leopard, Löwe, Bär, und viele andere Versuchsfahrzeuge und Panzerprojekte
by Podzun-Pallas Verlag.

We are interested in hearing from authors with book ideas on related topics.

Published by Schiffer Publishing Ltd.
4880 Lower Valley Road
Atglen, PA 19310
Phone: (610) 593-1777
FAX: (610) 593-2002
E-mail: Schifferbk@aol.com
Please write for a free catalog.
This book may be purchased from the publisher.
Please include $3.95 postage.
Try your bookstore first.

Foreword

The efforts to continue the development of the Panzer I and II, which had been regarded from the start as only transitory solutions, and in particular the development of an urgently needed armored scout car ("Luchs"), are portrayed, as well as the strivings for uniformity in Panzer III and IV on the one hand and V ("Panther") and VI ("Tiger") on the other.

The invasion of Britain that was discussed in the autumn of 1940 brought about intensive development of floating and diving tanks, which later gained importance again as the deep wading capability of the heavy tanks that exceeded the load limits of many bridges, and to this day is taken for granted as one of the requirements of a modern tank. A series of tanks was fitted with flamethrowers instead of their primary weapons, in order to give flamethrowers off-road capability and bring them into action with armored protection; other viewpoints were influential in the mounting of rocket launchers on tank chassis, which, to be sure, did not get beyond the project stage.

In view of the growing Allied air superiority, the call for effective Flakpanzer vehicles became more and more urgent and manifested itself in a number of such developments, many of which have already been portrayed in Volume 51 of this series. Several other interesting projects will be portrayed here.

The conclusion will consist of several drawings of heavy tanks about which little or nothing was previously known. They will serve as documents of what arose on the drawing boards of the designers without being able to be turned into reality.

The Neubaufahrzeug

In 1933 the OKH awarded a contract for the development of a new heavy tank based on the experience gained from the heavy tractor (1927-29). The designs for the Neubaufahrzeug (the project was carried out under this name) were to be built by the firms of Rheinmetall and Krupp.

The running gear of the tank resembled that of the VK 2001, also built by Rheinmetall. It consisted of five double-roller trucks and four return rollers that held the upper part of the tracks. The main turret was designed differently by Rheinmetall and Krupp. Krupp armed its tank with one 105 mm KwK and one co-axial 37 mm gun, while Rheinmetall planned on one 75 mm KwK with a 37 mm gun above it. Two machine guns, based on British (the "Independent" tank) or Soviet (T 32) models, were placed in separate, additional bow and stern turrets, identical to those of the Panzer I. In the inside of the tank, 80 rounds for the 75 or 105 mm tank gun and 50 rounds for the 37 mm gun could be carried, as well as 6,000 machine-gun bullets. The Neubaufahrzeug was armored relatively weakly, with 16 mm plate on the front and 13 mm on the sides. Because of this, the tank, based on its dimensions, had a low gross weight of 23 tons.

By 1935, a total of five prototypes of the Neubaufahrzeug had been built. Two of them had the Rheinmetall and three the Krupp turret. Since at that time there was no motor producing the required power available, six-cylinder BMW aircraft engines were used at first; later they were replaced by Maybach HL 108 motors that produced 280 horsepower. The two Rheinmetall vehicles, whose armor plate was made only of soft steel, were used for training purposes at Putlos. The three other tanks, with the Krupp turret, were shipped to Norway in April 1940, essentially only for demonstration purposes, to give the illusion that heavy tanks were on hand.

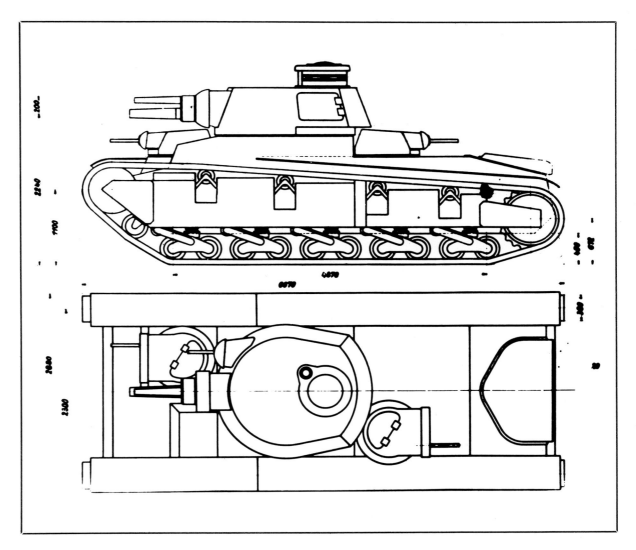

These two front views show the different arrangement of the weapons of the Krupp (left) and Rheinmetall (right) vehicles. The field of elevation for both vehicles ranged from -10 to +20 degrees. The 75 mm KwK L/24 gun used by Rheinmetall had a maximum range of two kilometers, while the 37 mm KwK L/46.5 used in both vehicles had a range of one kilometer. (2 x BA)

Right and below: These two pictures show the construction of the Neubaufahrzeug at the Rheinmetall factory. The builders are presumably working on the installation of the six-cylinder BMW aircraft engine, which produced 300 horsepower and gave the tank a top speed of 30 kph (1 x BA, 1 x IWM)

Below: In April 1940 all three Neubaufahrzeuge with the Krupp turret were shipped to Norway. Here one of these tanks is seen shortly after loading.

These three photos show the Neubaufahrzeuge in the harbor area of Oslo. In the picture at the upper right, the arrangement of the main turret and the two machine-gun turrets can be seen. The Neubaufahrzeug had ungreased link tracks which were driven by toothed wheels located at the rear, while in all other tanks the drive wheels were at the front. The crew consisted of six men. The radio equipment consisted of a twenty-watt UKW transmitter with a range of four to six kilometers and a similar receiver. (3 x BA)

This picture shows all three vehicles shortly after being shipped out. They were supposed to make it look as if large numbers of heavy tanks were available, and thus were often displayed. Because of their light armor plate (13 to 20 mm), though, the combat value of these vehicles was meager.

Upper left: After seeing action in the Oslo area, a Neubaufahrzeug supported an attack of four Panzer I and II tanks against British troops who landed at Andalsnes on April 17, 1940. This Neubaufahrzeug, though, had to be abandoned after it had gotten stuck. Here a Neubaufahrzeug rolls through a Norwegien town. The gun is protected from dust by a muzzle protector. Behind the tank is an Armored Command Car I Type B, of which in all only 200 were built .(BA)

Above and left: Here are two more pictures of these tanks in Norway. The two Neubaufahrzeuge that remained after the loss of one at Andalsnes were stationed at Akershus Fortress (Oslo) until the end of 1940. Later they are said to have taken part in the attack on Russia within Panzer Group I (von Kleist) and were lost near Dubno on June 28, 1941.

Aufklärungspanzer (Armored Scout Car) VK 601/1801

In 1939 the Daimler-Benz firm and Krauss Maffei were contracted by the HWA to develop the Panzer I further into both a light scout car (VK 601) and a heavy machine-gun carrier (VK 1801).

The first project was, as suited its intended use, relatively lightly armored (10-30 mm), so that its total weight was no more than eight tons, but it had a powerful Maybach motor that produced 150 HP. On the basis of these design features, the VK 601 had a high top speed of 65 kph. In addition, the designers laid out the vehicle so that it could also be used by the paratroops as an airborne tank. The armament was considerably better than that of the Panzer I. It consisted of a newly designed large-caliber machine gun (EW 141) and a coaxial MG 34 of the earlier type. In all, 46 of the VK 601 were built as an experimental series. Two of these were sent to Russia for troop testing (by the 1st Panzer Division) early in 1943, while the rest were reserves for the LVIII. Panzer Corps and took part in the fighting in Normandy in 1944.

In the development of the VK 1801, on the other hand, emphasis was placed on having the vehicle's armor plate as heavy as possible, not less than 80 millimeters, so as to offer protection from the anti-tank weapons of the time. The total weight of the ve-

One of the few action photos of the VK 601, probably taken during the Allied invasion of Normandy. The vehicle had a two-man crew. The driver was housed in the right side of the hull, and the commander sat directly behind him in the torret. He could observe his surroundings through eight periscopes (Kinon blocks) arranged in a circle in the cupola. (IWM)

hicle was therefore high, at 21 tons. Since the same motor was installed as in the VK 601, the vehicle was underpowered, with a power-to-weight ratio of 7.1 HP per ton, and it could only reach a top speed of 25 kph. The armament consisted of two MG 34 guns, both mounted in the turning turret.

Since such a support vehicle had already been left behind by other developments for the infantry, only 30 of the VK 1801 were built, all being delivered in 1940. Eight of them saw service along with the VK 601 with the 1st Panzer Division in Russia in 1943.

A VK 601 in the factory grounds of its builder, Kraus Maffei, still without any armaments. Because of its high top speed of 65 kph, self-lubricating tracks were planned at first (as can be seen in this picture). Later, though, they were replaced by tracks of the older type. The last vehicles in this series were fitted with a more powerful Maybach HL 61 motor and were designated VK 602. (BA)

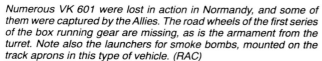

Numerous VK 601 were lost in action in Normandy, and some of them were captured by the Allies. The road wheels of the first series of the box running gear are missing, as is the armament from the turret. Note also the launchers for smoke bombs, mounted on the track aprons in this type of vehicle. (RAC)

Above: In order to keep the ground pressure of the 21-ton machine-gun carrier (VK 1801) as low as possible (it was 0.46 kg/sq. cm), the vehicles were equipped with what, compared to the Panzer I, were relatively broad tracks (0.54 m). Note also the unusual arrangement of the rounded entry hatch on the sides of the hull. (BA)

Upper right and right: In January 1942 driving tests in snow were undertaken with the VK 1801 at St. Johann in Tirol. In the lower picture, the tank is just being cleaned of ice and snow. The entry hatch and the armor plate of the roller shield can be seen. Of particular interest is the handhold above it, which could if necessary also be used as a step. The upper picture shows details of the running gear. This consisted, as in the VK 601, of five overlapping road wheels, sprung by transverse torsion bars. (2 x BA)

VK 901/1601

In 1938 the HWA called for a light, track-driven reconnaissance vehicle. For it, contracts were given to the Sauer firm of Vienna and to Daimler-Benz/MAN. While the Sauer firm projected a completely new wheeled/tracked vehicle, the combination of Daimler-Benz and MAN used the Panzer II, Type D as the basis for what was designated the VK 901.

In 1939 the first chassis of the VK 901 could already be delivered for testing. Instead of the required top speed of 60 kph, though, the vehicle could only reach 50 kph, since the Maybach motor's output of 145 HP was no longer sufficient for the vehicle's weight of 9.2 tons. The armament consisted of a 2 cm KwK 38 and a coaxial MG 34. In 1940 the production of a 0 series of 75 vehicles began, but after the twelfth vehicle, the production was halted, and none of the twelve saw active service.

In 1941 the VK 901 was modified into the VK 903, which differed from its forerunner in technical details. The newly developed Maybach HL 66, producing 200 HP, was to be installed, which would allow it to reach a top speed of 60 kph. But since this tank was already obsolete at this point in its development, only one prototype was ever built.

Running parallel to this development was that—also based on the Panzer II—of a "most heavily armored" type (VK 1601). In addition to a 0 series of 30 vehicles, which was built beginning in June 1940, the HWA also awarded a contract for 100 more of them, but the contract was withdrawn. Aside from the considerably heavier armor, the technical details (motor, etc.) were those of the VK 901, as was the armament. The relatively high fighting weight of 17 tons limited its top speed to 30 kph. No active service of the VK 1601 with the troops is known.

Below: The VK 901, developed from the Panzer II in 1938, was an attempt to fulfill the wishes of the HWA for a track-driven armored scout car. The thickness of the armor was 15 to 20 km, which allowed a light fighting weight of only 9.2 tons. Yet the required top speed of 60 kph was not attained.

1517.41

VK 1301 Luchs and other Reconnaissance Tanks

Above: Another concept of a track-driven reconnaissance tank was proposed by the Sauer firm of Vienna. Their newly developed behicle could run on either wheels or tracks. Accordingly, the top speed of the 6.5-ton tank was 80 kph with wheels or 30 kph with tracks. The turret of the armored scout car was almost identical to that of the VK 901. Since the production of the first prototypes lasted until June 1942 and the vehicle was outmoded by then, so no further development was done.

Opposite page, picture at right: The heavily armored development of the VK 901 was the VK 1601 (frontal armor: 80 mm, VK 901: 30 mm.). As in the VK 1801, the rounded entry hatches were set in the sides of the hull. The tracks, 0.23 m wide, were unlubricated and twice driven. The crew consisted of three men. The armament consisted of a 2 cm KwK 38 and an MG 34 mounted beside it. (BA)

Since all the wheeled vehicles previously used as armored scout cars had had major difficulties with the extremely unfavorable terrain in Russia, the need for a track-driven reconnaissance tank became more and more urgent. For that reason, the HWA ordered that the work on the VK 1301, which had been in progress since 1939 already, should be hastened. This vehicle's development had been based on the experience already gained from the VK 901 prototype. The tank was to correspond to the dimensions of the VK 901, but its fighting weight of 12.9 tons was some 2.7 tons more than that of the VK 901. In 1942 a prototype of the VK 1301 in soft-steel construction had been finished, and it was being tested exhaustively.

After a few minor improvements, the vehicle (VK 1303) went into series production as Panzer II, Tyle L (Luchs). The weight of the "Luchs" (Lynx) was one ton less than that of the VK 1301. The front armor plate was 30 mm thick, and the rest from 12 to 20 mm. The four-man crew had a 20 mm KwK 34 and a MG 34 mounted next to it as their armament. The 200 HP Maybach HL 66 engine already installed in the VK 903 gave the vehicle a top road speed of 60 kph.

All of the 100 "Luchse" that were built were utilized exclusively by reconnaissance units. There it soon became clear that the armor plate, like the armament with the 20 mm KwK, was inadequate. In order to improve the latter situation, a turret was built that was open at the top and carried the 50 mm KwK/L/60 formerly used in the Panzer III. But this version, of which 31 were delivered by the beginning of 1943, also fell short of fulfilling the needs of the troops.

The "Luchs", though, was only a maksehift solution anyway. The "Leopard" had been intended as the genuine reconnaissance tank. This decision was based on the experience gained with the VK 1601, which were evaluated in the VK 1602. Two versions of the "Leopard" were then developed: a light, thinly armored, fast-moving 18-ton type and a heavier one of 26 tons, which was to have armor plate of 20 to 60 mm (hull) and 50 to 80 mm (turret), and to be armed with a 50 mm KwK. While the troops pleaded for the lighter version, Hitler decided in favor of the heavier type in the summer of 1942. But then it was found that such a tank, aside from its armament, was very similar to the "Panther", and so the "Leopard" project was halted in January 1943, so that a reconnaissance version of the "Panther" could be developed instead. For this purpose, the HWA suggested that the turning turret with 50 mm KwK 39/L60, projected for the VK 1602 and later used in the eight-wheel "Puma" armored reconnaissance vehicle, be adopted. This project too was not followed up.

In the spring of 1943, after the production of the "Luchs" had been halted and the "Leopard" project had been abandoned, the Bohemian-Moravian Motor Works were commissioned to construct a transitional solution in the period from January to April 1944 on the chassis of the Czech Panzer 38 (t), with [p. 14] a new body, as a light reconnaissance tank. The armament to be installed was the 20 mm gun in a hanging mount, already used in the four-wheel Sd.Kfz. 222 scout car. Toward the end of 1944 the BMM was working on a scout car on the chassis of the "Hetzer" tank, which was to use a short 75 mm KwK on a fixed mount instead of the 20 mm hanging type. Series production was planned for mid-1945.

Above: A picture of the only completed VK 1301, including turning turret with full armament. (BA)

Left: The running gear of the VK 1301 was also subjected to thorough winter testing in St. Johann in Tirol in 1942. For these tests the vehicle was fitted with a temporary upper body that was to protect the crew from the cold a little. This body also enclosed the driver's seat, in front of which large windows with wipers were fitted in order to provide good visibility. (BA)

Ths VK 1301 charges over the deep snow of the terrain at top speed. The exhaust system of the tank was mounted at the rear, and can be seen here. (BA)

This picture shows the VK 1301 already fitted with the planned turret, but as yet without weapons. (BA)

Above: This view gives a good impression of the layout of the vehicle. The entire rear section was taken up by the engine, whose air-intake and ventilation openings, protected with grids, can be seen clearly. The body, unlike that of the VK 1301, extended out over the hull so it could hold the newly designed turret. (RAC)

Above: This side view from the rear clearly shows the exhaust system and the large entry hatch in the back wall of the turret. In addition, the crew could also get into the tank through a hatch on top of the turret. Appropriate to its use as a reconnaissance tank, the "Luchs" had a powerful radio, the umbrella antenna of which can be seen clearly here; it was also used in the "Panther" command tank. (BA)

Left: A side view of the Panzer II L ("Luchs"), later built in a small series. Unlike the VK 1301, the "Luchs" had full disc wheels. They had rubber tires and were mounted on transverse torsion bars. The first and last road wheels were also supported by one shock absorber each. (BA)

One of the extremely rare action photos of the "Luchs". It was used almost exclusively in Russia. The design of this tank had been made with the extreme climatic conditions of Russia in mind. Thus, it had a cooling-water exchanger that allowed it to be filled with already warmed cooling fluid from other vehicles when the outdoor temperatures were low. The armor plate, with a maximum thickness of 30 mm, was extremely meager. For extra protection, pieces of a Russian tank's tracks have been added above the driver's visor on the hull, just below the bottom of the turret. (BA)

By the end of 1943, a total of 100 Panzer II Type L, armed with a 20 mm KwK, had been delivered to the troops by MAN. Unlike the VK 1301, the main armament of the "Luchs" was not mounted to the right, but located in the middle of the cylindrical mantlet. The racks attached to this tank could hold reserve fuel canisters. With a fuel supply of 235 liters and a consumption of 150 liters per 100 km off the road, the "Luchs" had a range of 175 km. (IWM)

Left: A look inside the turret of the "Luchs". In the center is the 20 mm KwK, which was adjusted to the right elevation by handwheels. To the left of it are the attachments to which a MG 34 could be fitted if necessary. In all, the "Luchs" carried 330 rounds of 20 mm ammunition and 2,250 machine-gun bullets. (IWM)

Above: Probably the only surviving "Luchs" tank is seen today at the Royal Armored Corps Tank Museum in Bovington, England, along with numerous other German armored vehicles.

Upper and lower right: After "Luchs" production ended, but a real successor model had not yet been found, so the Bohemian-Moravian Motor Works created a makeshift solution early in 1944, using 118 chassis of the Panzer 38 (t) tank. The turret already used in the four-wheel armored scout car was used; it was open at the top and was fitted with a 20 mm KwK 38 L/55 beside a coaxial MG 42. The Praga EPA/2 engine gave the tank a top speed of 42 kph, and the armor plate was relatively thin, measuring 15 to 30 mm. Thus, this vehicle was inferior to the "Luchs" in every way. The Aufklärer 38 (t)—nowhere near all of which reached the troops—was used as of April 1944 by the armored reconnaissance units on both the eastern and western fronts. (2 x BA)

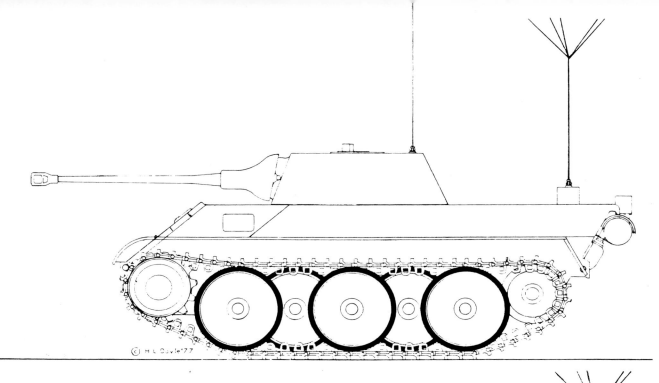

Since the "Luchs" tank was insufficient in both armor and armament, the MIAG firm was awarded a contract to design a successor model. For it, the VK 1601 was developed further into the VK 1602. This vehicle, also known as the "Leopard" reconnaissance tank, is shown in the upper drawing. But since there was a certain parallelism with the simultaneous development of the "Panther", the project was halted while still on the drawing board.

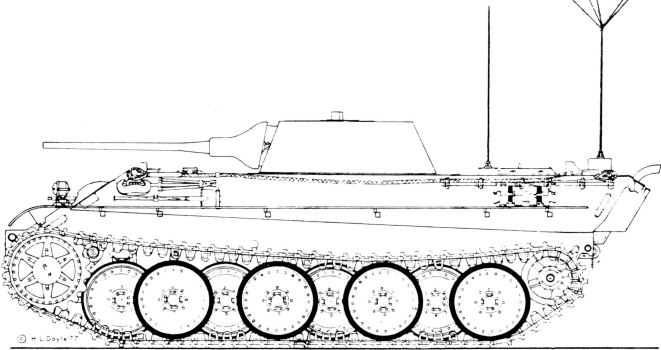

Instead of that, the possibility of designing a "Panther" reconnaissance tank was investigated (see lower drawing). This path too was not followed in the way illustrated, since the projected 50 mm KwK was regarded as an unduly weak armament. Instead, a prototype of the "Panther Armored Reconnaissance Car" was prepared. This vehicle was based on the "Panther" Type D but, instead of the 75 mm tank gun, was fitted with a dummy weapon.

Special Types of the Panzer III, IV and V VK 2001

When the In 6, under Generalmajor Lutz and Oberstleutnant Guderian, conceived the creation and equipping of the German armored troops in 1931-32, two tank types were projected, a lighter, faster battle tank with armor-piercing weapons (Panzer III) and a support tank with a large-caliber tank gun (Panzer IV). For the development of the latter, which was carried on until 1935 under the camouflage name of "Bataillonsführerwagen" (BW), and for which the HWA anticipated a weight of 18 tons, the firms of Rheinmetall and Krupp were given contracts.

The prototype built by Rheinmetall (VK 2001 RH) had a chassis of four trucks, each with two rubber-tired road wheels, very much like those of the "Neubaufahrzeug" developed by the same firm (see above). The tank had a 300 HP motor, which gave it a speed of 35 kph. But after several modifications, it was the Krupp design (VK 2001 K) that was put into series production later.

It should be noted that, in addition to the two firms named above, MAN was also invloved in the development of a similar 18-ton tank. This design (VK 2001 MAN), though, did not result in the building of a test vehicle.

Pz.Kpf.Wg. IV (7,5cm)

B.W. Holzmodell (Rh.)   K.772.A.36   z. Schreiben 1312 v. 26.2.35

The Panzerkam-pfwagen III/IV

In 1941 the HWA sought to achieve a certain uniformity in the Panzer III and IV, which would hopefully result in many advantages for their production. It was believed that, with the two vehicles built the same (except for their armament), production would be increased, repairs of damaged vehicles hastened, and training time shortened. In addition, the HWA was convinced that this would provide great flexibility in production, since the emphasis could be shifted to either one of the two types as needed.

A uniform box running gear was planned for the two vehicles, resembling that of the halftrack vehicles. The Famo firm had already gained experience in this realm and attained a decrease in roll and ground pressure, as well as a higher limit of wear. Several test vehicles with this new running gear were built on the basis of Panzer III and tested. Krupp was also working intensively on the Panzer III/IV.

Along with the further development of the previous Panzer III/IV, there were already numerous other designs based on the uniform chassis of Panzer III/IV. Some well-known types were the Sturmgeschütz III/IV assault gun, with a 75 mm L/70 gun, built by the Alkett and MIAG firms, a 105 mm assault howitzer, likewise made by Alkett, a heavy armored howitzer with its own ammunition transporter, made by the Stahlindustrie GmbH, and a Sturmpanzer III/IV.

The few Panzer III chassis with Famo box running gear were used almost exclusively for training purposes at various troop training bases. The lower picture shows practice in attaching a hollow charge to the engine compartment of a tank. The prototypes had the same turret as the Panzer III Type J, with the short 50 mm KwK L/42 (2 x IWM)

These efforts for uniformity were particularly urgent under the pressure of the conditions in the spring of 1944, yet the development of the Uniform Chassis III/IV was halted in October 1944, because there was no further demand for these obsolete tanks and their self-propelled mounts and other variants. Only the chassis of the Czech 38 (t) and the Panther/Tiger were to remain in use for armored scout cars, self-propelled gun mounts and assault guns.

The Panzer III with Famo running gear was also used for testing purposes, as seen here in the winter of 1943, to test the tank trailer sled built by the Ambi.-Budd.-Werke. This device was connected to the tank by a towbar three meters long. It was meant to provide effective infantry support for the tank in the winter conditions of Russia. This project too, sensibly, never went into production. (3 x BA)

In 1942 the Sauer Werke of Vienna developed a Panzer III railcar. A prototype was finished in mid-1943 and introduced to the HWA by the Arys troop training center in October of that year. The vehicle was meant to secure railroad lines in partisan areas, especially in Russia. To fit on the rails, the arrangement of the road wheels had to be modified from the usual pattern. Note the "buffers" on the front and back. (BA)

Left: Front view of the rail Panzer III. It had the turret of Panzer III Type N, with a 75 mm KwK L/24. The rail running gear, retractable by a lifting apparatus, was driven by the engine via four screw spindles. On test runs, speeds up to 100 kph were attained on railroad lines. Only a few prototypes of this vehicle were ever built. (BA)

Panther with Narrow Turret and Panther II

For the Panther, too, successor models were in the works by the end of the war. In 1944 Daimler-Benz was engaged in developing a new, reworked version (F) of the "Panther" tank, for which a number of small changes were planned, as well as a new turret, the so-called "Schmalturm" or narrow turret.

This was meant, in comparison with the old turret, to offer a smaller surface to be hit, and to be more heavily armored. Yet the space inside was not to be decreased, and a turret weight of eight tons was not to be exceeded. Instead of the roller mount used before, a "pig's-head" mount was planned, such as was used in the Tiger II, since it was much more likely to deflect shots. The turret was also prepared for the installation of the infrared night-vision device, which was then being developed.

As with Panzer III and IV, and for the same reasons (simplified production, easier service and repairs), the HWA required as much uniformity of Panzer V and VI ("Panther" and "Tiger") as possible. These modifications were designated Tiger II and Panther II. While series production of the Tiger II began at the end of 1943, only two prototypes of the Panther II had been built by the end of the war. The narrow turret already described was intended to be used on the Panther II. The installation of a recoil-free 88 mm KwK, as in the Tiger II, was investigated. In place of the box-type running gear of the Panther I, the "Staffel" running gear of the Tiger II was to be used, as it was considerably simpler and therefore could be built more quickly and cheaply. The fighting weight of the Panther II was about 50 tons, some seven tons heavier than the Panther I.

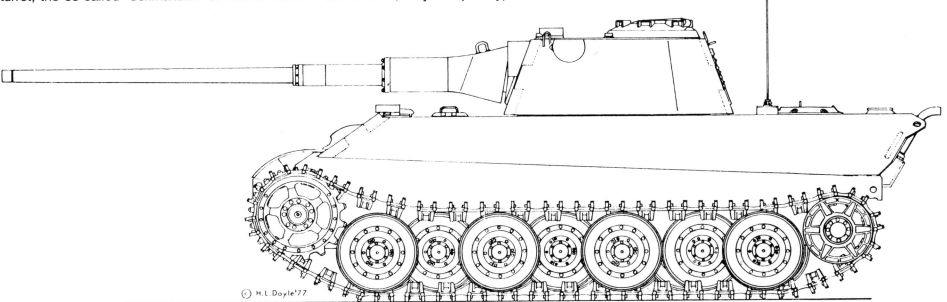

H.L.Doyle '77

A design drawing of the "Panther II", of which only two prototypes had been finished by the end of the war. One of them was taken to the U.S. Armored Troop School at Fort Knox, Kentucky, where it remains to this day.

Several Type G "Panthers" were fitted, for testing purposes, with the narrow turret planned for Type F.
The narrow turret was much more heavily armored than the earlier "Panther" turret; the front armor was 120 rather than the previous 80 mm thick, the side armor 60 instead of 45 mm. In place of the roller mount, it used the "pig's-head" mount, which deflected shots much better.

On the side and rear walls of the turret there were openings for machine pistols, since such an additional close-combat weapon had proved to be necessary, especially on the eastern front.
The narrow turret was also supposed to be used on the projected "Panther" II, where the possibility of mounting the 88 mm KwK used in the "Tiger" II, instead of the formerly usual 75 mm KwK, was to be explored. (2 x BA)

Flamethrowing Tanks

In order to make the flamethrowers used by the engineers for attacking buildings and bunkers more mobile, and especially to allow their use from a protected position, several tanks were equipped with a flame nozzle instead of their main weapon. These modifications were often undertaken by the troops themselves, using Panzer I to III, but some were also made by the industry.

In the Afrikakorps, Panzer I Type A tanks were fitted with a Flamethrower 40, which was installed in place of the righthand machine gun (shown in Volume 18, page 47). Ninety-five Panzer II, Types D and E, were also equipped with flamethrowers. The Flammpanzer III built on the basis of the Panzer III Type M was the only rebuilt tank that was produced in quantity by the industry. One hundred Panzer III tanks were supplied without weapons late in 1942 by the MIAG firm of Braunschweig to the Wegmann AG in Kassel, which equipped the tanks with the flamethrowing apparatus. In place of the 50 mm KwK they had a 14 mm flamethrowing pipe. The container that held the burning fluid was located inside the tank. The flamethrowing tanks were meant to be used in the built-up area of Stalingrad.

In December 1942 the first two companies could be supplied with these vehicles, but because of the war's events, the tanks never reached their original area of use. Since the vehicles also failed to meet the troops' requirements, a number of them were soon rebuilt into normal battel tanks. Despite these negative experiences with the Flammpanzer III, Hitler ordered in November 1944 that the flamethrowing apparatus left over from rebuilding the Flammpanzer III be installed in available tanks. In all, 35 Panzer III (with two flamethrowers in their turrets), Sturmgeschütz III and Jagdpanzer 38 (t) "Hetzer" (with the gun replaced by a flamethrowing pipe) were rebuilt. Hitler even wanted the Jagdtiger to be rebuilt into a flamethrowing tank, but his wish was not fulfilled.

This factory photo shows a newly delivered Flammpanzer III. One hundred Panzer III Type M were fitted with a flamethrowing apparatus by the Wegmann firm of Kassel. (BA)

Some of the heavy "Char B1" tanks captured in France were also rebuilt and used as flamethrowing tanks. The 75 mm howitzer at the front of the tank was replaced by a flamethrower. (BA)

Ninety-five Panzer II Type D and E tanks were each fitted with two flamethrowers, one mounted in the gun mount and the other in a separate turret at the front of the tank. This apparatus, though, has a meager range of 25 meters. (BA)

In November 1944, Hitler ordered that available flamethrowing apparatus be mounted to available tanks in place of their primary armament. This process included the Sturmgeschütz III. A Sturmgeschütz thus modified is shown in this picture. The exact number of rebuildings of this kind, though, is not known. (BA)

The flamethrowing tank most widely used by the troops was the Flammpanzer III. It carried 1,020 liters of inflammable oil in two containers, which were installed inside the tank. The three smoke-cartridge launchers attached to the turret walls can be seen clearly here. Almost every armored unit had a platoon with five flamethrowing tanks. (BA)

Two Flammpanzer III at a troop training center. The four-digit numbers on the turret can no longer be identified. The burning oil carried inside the tank was directed into the pipe by a pump and there ignited by high-tension electricity (1,000 V, 300 A). The flamethrower was operated, via foot pedals, by the tank commander, who also had to operate the machine gun, elevation apparatus and turret turning apparatus. The flamethrowing pipe had a field of elevation from +20 to -10 degrees. Thus, targets of varying heights and ranges could be attacked. After every burst of fire, though, the air was so full of smoke that the day was turned to night and new targets could scarcely be seen. (BA)

Here a Flammpanzer III of the 44th Infantry Division is seen. The burning oil capacity of 1,020 liters allowed eighty bursts of flame, each two to three seconds long. These vehicles, though, did not live up to the troops' expectations. For that reason, some of the Flammpanzer III tanks were soon (1943) rebuilt into normal battle tanks.

Rocket Launchers

Toward the end of the war, the possibility of installing launching frames for rockets on tank chassis, in addition to their customary use on the Schützenpanzerwagen, so that the launchers could be used in rough country and from protected positions. Thus, a number of captured French UE infantry tractors were fitted with 28/32 cm launching frames. These were mounted on a platform, on the rear of the vehicle, which could be raised and lowered depending on what shot angle was wanted. A rocket launcher was also mounted on the Panzer IV chassis in place of the turning turret. Prototypes of this vehicle were produced and used.

The Skoda firm also worked on a project in which the 105 mm rocket launcher was mounted on the "Panther" tank chassis. The launching chambers were attached to the mount of the 88 mm Flak gun. This was mounted in place of the turret and could be turned. Because of the constant shortage of "Panther" chassis, this design was not put into production.

Right: Toward the end of the war, the Skoda firm designed a 105 mm rocket launcher that was to be mounted on the chassis of the "Panther" tank. The range of elevation of the launcher extended from -5 to +75 degrees. The chambers, 3.5 meters long, were welded together of steel angle irons.

Right: This picture shows the rocket launcher on a Panzer IV chassis. In front of the launching frame, which can be tilted at various angles by a hydraulic lift, is an armored cabin with a machine gun for the crew's use. The bow machine gun usually mounted on the Panzer IV has been replaced by a visor slit.

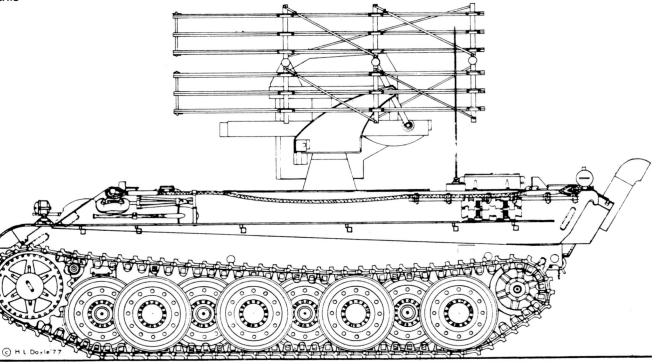

Diving and Swimming Tanks

After the end of the French campaign in the summer of 1940, Germany developed plans for crossing the English Channel and invading Britain (Operation "Sealion"). To do this, there was a need for amphibian and diving armored vehicles. In October 1940, an armored unit of volunteers was assembled at Putlos in October 1940 to be trained for the projected invasion of Britain. This unit was equipped with suitably modified Panzer II and III tanks.

The Waprüf 6 had thus ordered floating units which could be disembarked from landing craft, and would be stable in a Strength 3 to 4 seaway, from the Sachsenberg firm of Roslau. In all, fifty-two of these floating units, which were held by the return rollers of the vehicles, were delivered to Putlos. The Panzer Ii tanks so equipped were driven in the water by two ships' propellers, which were linked to the drive wheels of the tank by a driveshaft. Since the tank would only be immersed in water up to the level of the track aprons, it was able to fight while in the water.

The Panzer III was to be landed in a very different manner. The tanks were to be equipped for diving to a depth of 15 meters, so that they could reach the beach by driving on the sea bed. For this, a complete waterproofing of the tank was necessary, and this was done with cable tar. The air intake over the engine compartment was also closed completely, and the weapons were covered with rubber coatings to prevent water from entering them. The fresh air for the crew and the engine was supplied from the surface by a hose 18 meters long. On the end of the hose that projected out of the water there was a radio antenna that allowed the tank to be directed from the landing craft. The engine exhaust gases were to be discharged directly into the water. Thus, the exhaust pipes were fitted with one-way valves to prevent water from entering. During thorough tests, though, it was found that this system functioned only at shallow depths, and that the conduction of the exhaust to the surface at a depth of 15 meters was impractical.

When the Russian campaign drew near, the Tauchpanzer III was remembered, and the tanks were re-equipped for fording rivers. For this, the rubber hoses for the air intake were replaced by steel pipes 3.5 meters long. These Tauchpanzer III tanks were used by the 18th Panzer Regiment when they crossed the Bug on June 22, 1941.

The planning of Undertaking "Sealion" had inspired several German firms to begin designing genuine swimming and diving tanks. For example, an underwater diving vehicle (UT-Kampfwagen), which could operate like a submarine two to six meters below the surface, was developed. By means of trimming tanks mounted on the sides of the tank, the height of the vehicle could be controlled, and in addition, they could be used to equalize possible differences in attitude by appropriate flooding.

A different project was the "Krokodil" tank developed by Krupp in 1942. The vehicle was meant to be able to cross waterways up to twelve meters deep and one kilometer wide. Its weight was projected at 28 tons. The power unit for underwater driving was debatable: either a 100 HP battery-operated electric motor or a Diesel engine. Experience gained in submarines was to be evaluated for this purpose. A further project was the "Schildkröte" (Turtle) amphibian armored scout car developed by Trippel on the basis of its SG 6 amphibian; three prototypes of it had been built by 1942.

A different amphibian vehicle had been developed even before the war began, by the Alkett firm in co-operation with the Boitzenburger Binnenwerft; this was the Land-Wasser-Schlepper (LWS), of which 21 were built. The LSW was to be used, as seen by the OKH, as a towing tractor on the land and in shallow water, as well as a motorized towboat in the water itself, particularly for landing craft. These vehicles had a boat-shaped hull that was mounted on a stretched tank chassis.

In 1941, Magirus developed the LWS into the so-called Panzerfähre, using a Panzer IV chassis. When a deck was suspended between two such vehicles, one could use it to ferry tanks up to the size of the Panzer IV. Unlike the LWS, the Magirus SW+MK was safely armored. By May 1942, two prototypes had been built, and they were then tested thoroughly. Meanwhile, though, the fighting weight of the tanks had risen and the carrying capacity of the armored ferry was no longer sufficient. Thus, this project was not developed any further.

Carrying-capacity problems of another kind occurred in the production of the "Tiger" tank, also begun in 1942, as it was too heavy for many bridges. Thus, the first 495 Tiger I tanks produced were equipped with [p. 35] a wading apparatus, which resembled that of the Panzer III described above and allowed the fording of rivers up to four meters deep. The air intake system consisted of a 4.5-meter stack on the back of the vehicle; the exhaust gases were discharged directly into the water. The engine compartment was carefully sealed off from the fighting compartment to prevent the poisonous exhaust gases from penetrating. Because of the high cost, the wading apparatus was later eliminated from the "Tiger", but it was projected again for the overweight "Maus" tank.

Above: Front and rear views of the Schwimmpanzer II while being tested before entering the water. The floats attached to the lengthened return-roller axles were developed by the Sachsenberg firm of Roslau, had three chambers each, and were filled with small celluloid pipes. Originally, tanks made seaworthy in this manner were intended for use in the invasion of Britain. But when the German leadership changed its plans and gave up Operation "Sealion", the floats were removed and the vehicles used on the eastern front as normal Panzer II tanks.

Right: The Schwimmpanzer II during testing. As can be seen, only the chassis was under water. Thus, the turret was always able to fight. The rear area of the tank was covered with a sheet-metal deck and the ventilators for the engine and such were located there and protected from being entered by water. The exhaust discharge also presented no difficulties.

Upper left: On July 9, 1940, the Tauchpanzer III was displayed to Army and Navy officers near Wilhelmshaven. The 15-meter rubber hose, reinforced with iron rings, that provided fresh air can be seen, as can the watertight coverings of the weapons. Explosive fuses were placed in the coatings, so that after coming out of the water, the crew could blow off the protective covering to make the tank ready for action in the shortest possible time. (BA)

Above: Except for the measuring poles attached for the test run, nothing more of the submerged Panzer III can be seen. In the foreground is the buoy to which the air hose is attached, with the radio antenna by which the tank receives its steering commands. (BA)

Left: After all watertight coatings have been checked, the Panzer III is lowered into the water by a crane. It was planned that the vehicles were to get into the water on a ramp from their ship to go into action. (BA)

Above: In 1936 the firm of Rheinmetall-Borsig had already been directed to collaborate with other firms in developing an amphibian towing tractor, also called a land-water tractor. This rear view clearly shows the two screw propellers and the rudders. The engine exhaust was ducted off via two separate exhaust systems in the rear of the hull.

Upper and lower right: All Schwimmpanzer II and Tauchpanzer III had been assigned to Panzer Regiment 18 and remained with it after the plans to invade Britain had been given up. In the spring of 1941, the regiment was given the task of checking all the waterproofing on their Tauchpanzer III to determine their readiness for service. Instead of the previously used long rubber hoses (see upper right), they were now fitted with 3.5-meter steel pipes for wading rivers. After test runs in the Werbellin Lake, the tanks saw action crossing the Bug on June 22, 1941.

Side and front views of the Land-Wasser-Schlepper, which carried a crew of twenty men. The cooling air for the Diesel engine was discharged through the rear shaft of the two that are visible on the roof. The front shaft is for ventilation of the interior and has a 9.3 HP ventilator fan. On the sides, pieces of equipment such as ropes are attached. Seven prototypes were built; they had larger front panels than the later vehicles, and they were tested thoroughly in 1940. Later fourteen more Land-Wasser-Schlepper were built. The pictures show this later vehicle. (3 x BA)

Some of the 21 Land-Wasser-Schlepper were assigned to the shipyard companies of the landing engineer battalions. The vehicles were used primarily in the eastern theater of war, such as in the occupation of the Baltic islands in 1941. The LWS was to be used as a towing tractor on land and a motorized towboat on the water. For these tasks, the vehicles were equipped with a powerful winch. (BA)

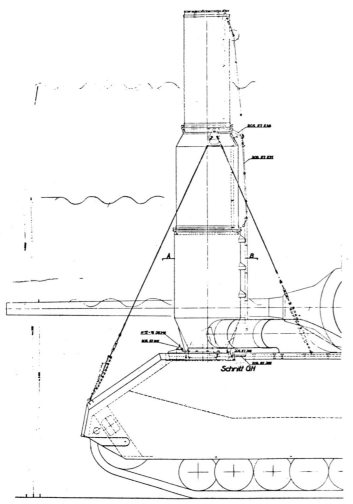

Schnitt GH

Since the load limits of many bridges were insufficient to carry the 56-ton "Tiger" I, the first production run was equipped with a wading system, which allowed them to ford waters up to 4.1 meters deep. On the grounds of the Henschel factory in Haustenbeck, extensive diving tests were carried out with the "Tiger" in 1942 and 1943. It was determined that the tank could spend up to two and a half hours under water with its engine running. In this picture, the ventilating pipe attached to the rear body can be seen very well. (BA)

Upper left: The further development of the unarmored Land-Wasser-Schlepper resulted in the two prototypes of an armored ferry built by Magirus and tested in the summer of 1942. Propulsion in the water was done by the main engine (Maybach HL 120), and the land or water drive train was chosen by the gearbox. To cool the engine, air ejection shafts were attached to the top of the vehicle.

Left: A wading apparatus for water depths up to eight meters was also planned for the 200-ton "Maus" tank. It was intended to allow the tank to cross waterways under its own power (see drawing). It was soon recognized, though, that the cooling and exhaust problems of the Diesel engine could not be overcome without great trouble. It was then considered that the electric motors of the Diesel-electric powerplant of the "Maus" might be powered through cables from the Diesel engine of a second tank on the shore for underwater movement.

Flakpanzer Projects

In 1942 it had already become clear that the armored units, exposed to fighter-bomber attacks, could only be protected from this danger by mobile Flak units. But since the development of a suitable Flakpanzer would take some time, a transitory solution was found by mounting quadruple 20 mm and single 37 mm Flak guns on Panzer 38 (t) and Panzer IV chassis. At first the weapons were merely surrounded by side walls that had to be folded down in action (Flakpanzer "Möbelwagen"); later the guns were installed in open-top turning turrets (Flakpanzer "Ostwind" and "Wirbelwind"). These vehicles all went into series production and have been described in depth in Volume 51. Along with them, though, there also existed a number of interesting project studies that will be examined somewhat more closely in the illustrations that follow.

What was wanted as a final solution was a Flakpanzer with a closed gun turret. Only two designs of this type, the "Kugelblitz" and "Coelian" Flakpanzers, developed beyond the drawing-board stage to the building of prototypes or of a wooden model in 1:1 scale.

The "Kugelblitz" Flakpanzer had been developed since 1944 at the Daimler-Benz factory in Berlin-Marienfelde. The chassis used was that of the Panzer IV. The Flak turret was a new design that pointed the way to the postwar era. The intended weapon was a twin 30 mm AA gun that had been developed by the Rheinmetall firm from the 30 mm Flak MK 103. The most modern feature of this weapon was the belt feed for the ammunition. It made this gun the first belt-feed type in the German Army, giving it a cadence of 425 rounds per minute with a range of 5,700 meters.

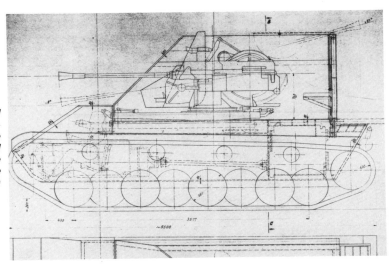

Typical of the many designs of self-propelled armored gun mounts with light anti-aircraft guns was this design by the Krupp firm. According to factory data, the Device 338 V4 was planned as its weapon. The hull has angled armor plates, like the "Panther" tank. The chassis of the vehicle shows similarities with the Panzer IV, though the road wheels overlap here. (BA)

The twin guns were mounted, fixed in place, in a closed "skullcap" housing. The "skullcap" was anchored in a stumpy protective cover and had an elevation range of -7 to +80 degrees. A hydraulic elevating apparatus allowed an aiming speed of 60 degrees per minute. The commander and aiming gunner both sat in the turret. At the end of 1944, five more prototypes of the "Kugelblitz" were produced. Series production was to begin in February 1945, but the events of the war prevented this.

Since the beginning of Flakpanzer development in 1943, there had also been discussion of the possible use of the "Panther" chassis as a carrier for anti-aircraft guns. The installation of weapons of various calibers (37, 55, 88 mm) was tested. While the 55 and 88 mm Flak guns were to be installed in open-top turrets similar to that of the "Wirbelwind", the twin 37 mm guns of the "Coelian" Flakpanzer were to be installed in a closed turret. The firms of Krupp and Rheinmetall-Borsig collaborated on the design of the "Coelian". To be sure, the further development of the vehicle, after a "Panther" Type D had been fitted with an appropriate dummy turret, was halted at the end of 1944, since the Inspector General of the Panzer Troops had regarded the firepower of the 37 mm gun as too meager compared to the weight of the vehicle. Instead, the further development of a Flakpanzer on the "Panther" chassis, which had been in progress since mid-1944, was urged. It was to carry twin 55 mm guns—likewise in a closed turning turret—but the work could not be finished before the war ended.

G11504/2

STAATSGEHEIMNIS lt. §88R STGB

G11504/3

14536 MT

STAATSGEHEIMNIS lt. §88R STGB

STAATSGEHEIMNIS lt. §88R STGB

These three pictures show the Rheinmetall-Borsig firm's attempt to mount the 37 mm Flak gun in an open-top turning turret, on a chassis, lengthened by two road wheels, of the "Luchs" reconnaissance tank. Since the hull of the "Luchs" was considerably narrower than that of the Panzer IV, the turret projects over the tracks at the sides. More detailed information about this vehicle has not been found. (3 x BA)

This wooden model represents the "Kugelblitz" Flakpanzer, of which five prototypes had been built by the German Iron Works in Duisburg before the war ended. At the end of 1944, they, along with several "Möbelwagen", "Wirbelwind" and "Ostwind" Flakpanzers, were assigned to a newly organized Armored Flak Replacement and Instructional Unit at the Ohrdruf troop training center in Thuringia. The "Kugelblitz" had a very low silhouette; its height was only 2.3 meters. The 30 mm double gun was mounted in a rounded "skullcap" housing. The "Kugelblitz" Flakpanzer was a successful design in many ways, pointing the way to the Flakpanzer development of the fifties . (2 x BA)

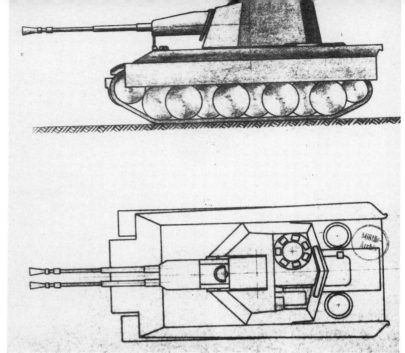

Left: In 1943, the development of the "Coelian" Flakpanzer was already underway. It was to have twin 37 mm guns in a closed turning turret. Designs for the turret were made by the firms of Daimler-Benz and Krupp. Early in 1944 the Krupp design was chosen. In the same year, the firm of Rheinmetall finished a mockup of the turret, which was mounted on a Type D "Panther" chassis for testing. The 37 mm twin Flak guns were to have a cadence of 2 x 500 rounds per minute. (2 x BA)

In mid-1944 the Krupp firm was instructed to undertake the installation of twin 55 mm Flak guns, instead of the 37 mm guns, on a "Panther" chassis. The drawing above shows side and top views of one suggested possibility. The tank was to have a four-man crew and an ammunition supply of 104 rounds. An elevation range of -5 to +80 degrees was planned for the 55 mm 58 Rh.D.V. 3 twin guns. The turret weight was estimated at 9,060 kilograms. (BA)

Heavy Tank Projects

In the spring of 1942, the Krupp firm was working on the design of another heavy tank, which was designated "Löwe" (Lion) or Panzer VII and weighed more than the "Tiger". It was based on the VK 7001, for which a lighter (100 mm front armor) 76-ton and a heavier (120 mm front) 90-ton version were envisioned. A 105 mm L/70 KwK and a turret machine gun was planned for both. The crew was to number five men, the top speed was to be 26.8 kph (76-ton type) or 23 kph (90-ton type).

At that time, Hitler was already thinking in terms of as heavy armor and as large a caliber as possible,

at the cost of speed. Thus, the lighter version was dropped and the 90-ton tank was modified to carry a 150 mm KwK L/40 or L/37 gun, and front armor plate of 140 mm. The top speed was to be increased to 30 kph, and the ground pressure to a maximum of 1 kg/ sq. cm, by increasing the track length by 4.96 meters and the track width to 900 or even 1,000 mm. Meanwhile, though, the decision had been made in favor of an even heavier tank weighing over 100 tons, the "Maus", for which Krupp provided a series of designs which were later rejected in favor of the Porsche design. It is interesting, though, that in June 1942 Oberst

Fichtner (HWA) suggested a shortened 80-ton version of the "Löwe" with an 88 mm KwK L/71, 140 mm front armor, and a top speed of 35 kph. Its performance data were amazingly like those of the later "Tiger II".

This design drawing shows the first version of the 90-ton "Löwe" tank with 105 mm gun. This drawing and the following one (front view of the Löwe) are reproduced here for, we believe, the first time.

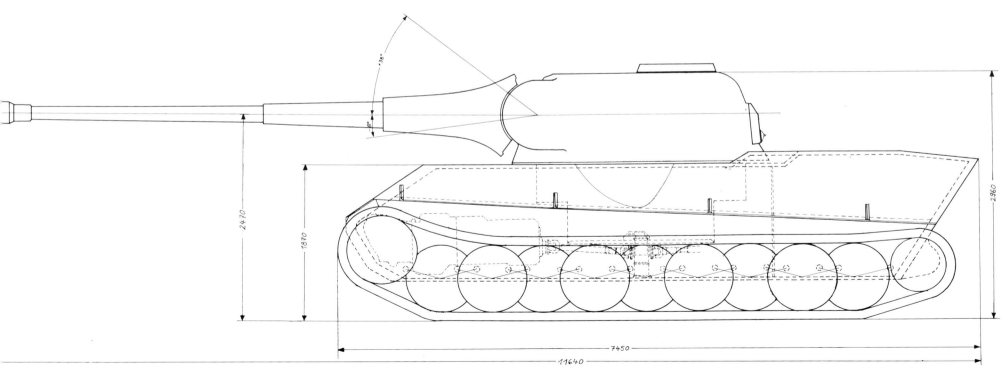

The design of an armored 120-ton vehicle with a 305 mm L/16 gun and known as "Bär" (Bear) is dated May 10, 1943. It was to be driven by a 700 HP Maybach HL 230 engine and reach a speed of about 20 kph. The running gear was largely taken from the "Tiger". The rigidly mounted gun had an elevation range of 70 degrees and was to fire explosive shells weighing 350 kg (50 kg charge) 10.5 kilometers at a muzzle velocity of 355 meters per second. The "Bär" was never built (See inside back cover for interior drawing).

Probably the most unusual tank project was proposed by Dipl.-Ing. Grote, in charge of U-boat construction in Speer's ministry, in June 1942: a tank weighing approximately 1,000 tons in all (Project P 1000). The vehicle, 35 meters long and 14 meters wide, was to run on two wide track systems, each with tracks 3.5 meters wide. Two MAN Diesel engines of 8,500 HP each, or eight Daimler-Benz powerboat motors of 2,000 HP each, were to provide a top speed of 40 kph. The proposed weapons were 280 and 128 mm ships' guns plus eight 20 mm Flak or HD 151 guns in turning-ring mounts. In December 1942, Krupp offered an alternative design with somewhat different goals in mind, a vehicle of some 1,500 tons with 250 mm front armor, carrying an 800 mm gun and driven by two Diesel U-boat engines. Fortunately, it was soon realized that such monsters were a senseless waste of materials and the development was halted. Designs of these monsters have not been known before.

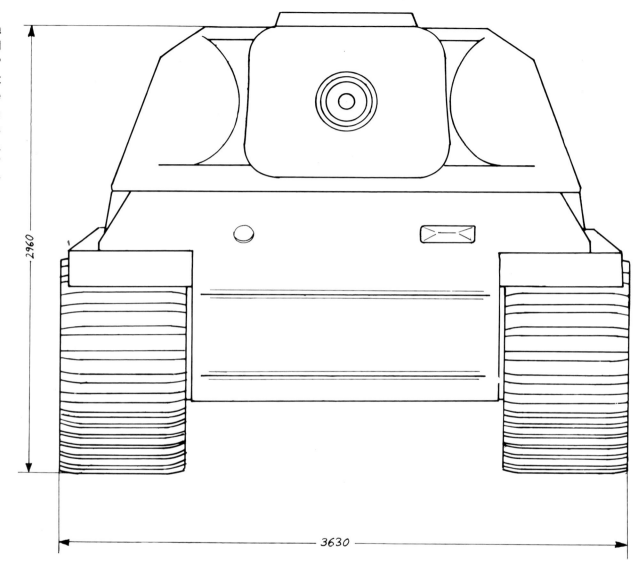

Technical Data

Type	Neubaufahrzeug	VK 601	VK 1801
Manufacturer	Rheinmetall(Krupp)	Krauss-Maffei	Krauss-Maffei
Year built	1935	1919-1941	1939-1940
Crew	6 men	2 men	2 men
Weight	23 tons	8 tons	21 tons
Length	6.65 meters	4.20 meters	4.37 meters
Width	2.90 meters	1.92 meters	2.64 meters
Height	2.90 meters	2.01 meters	2.05 meters
Engine	BMW 6-cyl aircraft, later Maybach V-12 HL 108 TR	in-line 6-cyl. Maybach HL45p	in-line 6-cyl. Maybach HL 45p
Cooling	Water	Water	Water
Power	290 HP	150 HP	150 HP
Transmission	6 fwd, 1 reverse	8 fwd., 1 rev.	4 fwd., 1 rev.
Steering	Wilson planetary	Three-wheel KM LG45R	Clutch
Clearance	0.45 meter	0.29 meter	0.35 meter
Front armor	20 mm	30 mm	80 mm
Side armor	13 mm	10 mm	80 mm
Rear armor	13 mm	10 mm	80 mm
Armament	75 mm L/23, 5 KwK (Rheinmetall), or 105 mm KwK (Krupp) + 37 mm L/45 KwK + 3 MG	EW 141 KwK	2 MG 34
Running gear	2 118-link tracks, rear drive, idler, 10 road wheels, 4 return road wheels	2 89-link trks. front front drive, rear idler, 5 no road wheels, no rollers	2 53-link trks. front drive, rear idler, 5 return rollers
Track width	0.38 meter	0.39 meter	0.54 meter
Fuel capacity	?	?	?
" consumption	?	?	?
Top speed	30 kph	65 kph	25 kph
Ground pres.	0.69 kg/sq. cm	0.84 kg/sq. cm	0.46 kg/sq. cm

Type	VK 901	VK 1601	Panzer II L 38(t) "Luchs"	Scout Car with 20 mm KwK
Made by	MAN	MAN	MAN	BMM
Yrs. built	1941-42	1942	1942-43	1944
Crew	3 men	3 men	4 men	4 men
Weight	10.5 tons	18 tons	11.8 tons	9.75 tons
Length	4.24 meters	?	4.63 meters	4.51 meters
Width	2.38 meters	?	2.48 meters	2.14 meters
Height	2.05 meters	?	2.21 meters	2.17 meters
Engine	in-line 6-cylinder HL 66 p	HL 45 p	Maybach HL 66 p	Praga EPA/2
Cooling	Water	Water	Water	[not listed]
Power	200 HP	150 HP	200 HP	?
Gearbox	5 fwd/1 rev.	5 fwd/1 rev	6 fwd/1 rev.	5 fwd/1 rev.
Steering	?	?	MAN clutch	?
Clearance	?	?	0.40 meter	?
Frt. armor	80 mm	80 mm	30 mm	50 mm
Side armor	15 mm	50 mm	20 mm	15 mm
Rear armor	15 mm	25 mm	20 mm	15 mm
Armament	EW 141	KwK 20 mm KwK 38 L/55	KwK 20 mm 38 L/55	KwK 20 mm 38 L/55
	+ 1 MG 34	+ 1 MG 42	+ 1 MG 34	+ 1 MG 42
Runn. gear	2 tracks front drive rear idler 5 overlapp. road wheels no return rollers	2 tracks front drive rear idler 5 overlapp. road wheels no return rollers	2 tracks front drive rear idler 5 overlapp. road wheels no return rollers	2 tracks front drive rear idler 4 road wheels no return rollers
Track width	?	?	0.36 meter	?
Fuel capacity	?	?	235 liters	?
Consumption	?	?	150 l/100 kph	?
Top speed	50 kph	31 kph	60 kph	42 kph
Pressure	?	?	0.98 kg/sq.cm	?
Climbing	?	?	30 degrees	?
Turning	?	?	In place	?
Suspension	?	?	Transverse torsion bars	?
Climbing	?	30 degrees	?	?
Turning circ.	?	In place	In place	?
Suspension	?	?	Transverse torsion bars	?

Type	Flammpanzer III	"Kugelblitz"lt. Flakpanzer
Manufacturer	MIAG	Daimler-Benz
Year built	1942	1945
Crew	3 men	5 men
Weight	23 tons	25 tons
Length	6.40 meters	5.92 meters
Width	2.97 meters	2.95 meters
Height	2.50 meters	2.30 meters
Engine	Maybach V-12 HL 120 TRM	Maybach V-12 HL 120 TRM
Cooling	Water	Water
Power	300 HP	300 HP
Gearbox	6 forward speeds, 1 reverse	6 forward speeds, 1 reverse
Steering	Daimler-Bz./Wilson	Krupp/Wilson
Ground clearance	0.38 meter	0.40 meter
Front armor	80 mm	30 mm
Side armor	30 mm	30 mm
Rear armor	50 mm	50 mm
Armament	14 mm flamethrower + 2 MG 34	twin 3 cm Flak MK 103/38 + 1 MG 42
Running gear	2 93-link tracks rear drive, front idler, 6 road whls, 3 return rollers	2 99-link tracks rear drive, front idler, 8 road wheels, 4 return rollers
Track width	0.40 meter	0.40 meter
Fuel capacity	320 liters	470 liters
Fuel consumption	182 liters/100 km	220 liters/100 km
Top speed	40 kph	38 kph
Ground pressure	1.03 kg/sq. cm	0.89 kg/sq. cm
Climbing ability	30 degrees	30 degrees
Turning circle	5.85 meters	5.92 meters
Suspension	Transverse torsion	Transverse torsion

Type	VK 7001 "Löwe"	"Bär" Tank
Manufacturer	Krupp	Krupp
Year built	1942	1943
Crew	5 men	6 men
Weight	90 tons	120 tons
Length	7.74 meters - gun	8.20 meters
Width	3.83 meters	4.10 meters
Height	3.08 meters	?
Engine	Maybach V-12 HL 230 P 30	Maybach V-12 HL 230 P 30
Cooling	Water	Water
Power	800 HP/1000 rpm	700 HP/3000 rpm
Gearbox	12 speeds	6 forward, 1 reverse
Steering	?	?
Ground clearance	0.50 meter	?
Front armor	120 mm	130 mm
Side armor	100 mm	80 mm
Rear armor	?	?
Armament	150 mm L/40 KwK	305 mm L/16 KwK
Running gear	2 ?-link tracks front drive, rear idler, 9 overlap. road wheels, no return rollers	2 ?-link tracks front drive, rear idler, 10 overlap. road wheels, no return rollers
Track width	0.90 meter	1.00 meter
Fuel capacity	?	?
Fuel consumption	?	?
Top speed	23 kph	20 kph
Ground pressure	1 kg/sq. cm	1.13 kg/sq. cm
Climbing ability	?	?
Turning circle	?	?
Suspension	Torsion bars	Leaf springs

Festivals_Milestones_Young People and God | Nick Harding

DIY
Celebrations

Acknowledgements

Text copyright © 2000 CPAS

CD copyright © 2000 CPAS
Resource format and concept © CPAS
This edition copyright © 2000 CPAS

Published by
CPAS
Athena Drive
Tachbrook Park
WARWICK
CV34 6NG

ISBN 1 902041 08 9

Printed material
Purchase of *DIY Celebrations* permits the photocopying of all printed material for use with the churches or groups for which the purchaser has responsibility, and for training other leaders in those churches and groups. If *DIY Celebrations* has been purchased by a church or group, for the purposes of the above 'the purchaser' shall mean the current minister or leader. The photocopiable visuals (pages 12-17) may be copied and adapted freely.

The Bible text is from the *Holy Bible, New International Version* copyright © 1973, 1978, 1984 by International Bible Society.

CD
Electronic text and illustrations
The text and illustrations of *DIY Celebrations* on the CD are copyright and all rights are reserved. Ownership of the CD entitles the purchaser to view and copy the material for use within the purchaser's own church group.

The text is only licensed for a single PC and may not be placed on any form of computer network without prior written permission from CPAS.

Copies may be kept for backup purposes only.

Other material on the CD may be freely used by the purchaser as above but may not be copied by any means without prior written permission from CPAS.

Additional material and software placed on the CD by CPAS are copyright according to the details distributed with the material. CPAS is not responsible in any way for the satisfactory use of this additional material which is provided on an 'as is' basis. All trade marks are acknowledged.

Music
With thanks to Alliance Music; Wildgoose Productions; Julia Kelly

CD origination by Vernon Blackmore
CD production by Copymaster (UK) Ltd

Written by Nick Harding
Additional material by David Bell
CPAS is grateful for invaluable help and advice from Elinor Brien, Terry Clutterham, Geoff Harley-Mason, Vandella King, Philip Mounstephen and Steve Tilley.

Edited by AD Publishing Services Ltd
Design by ie Design
Illustrated by Gary Dillon
Printed by Unigraph Printing Services

Church Pastoral Aid Society
Registered Charity No 1007820
A company limited by guarantee

British Library Cataloguing-in-Publication Data. A catalogue record for this resource is available from the British Library.

Introduction

Welcome to DIY Celebrations

It's great when you have something to celebrate. And we do. There's all that God has done for us in Jesus. Worth a party any day. And there are all the ways that God shows his love, care, strength and healing to us through his Holy Spirit in our everyday lives, too. And all that includes 11- to 14-year-olds, of course. Which is the reason for this resource.

DIY Celebrations has a large number of ideas for marking different events through the year. Some of these are established occasions such as Christmas and Easter. Others are big events in the lives of young people, such as moving home. And others are significant times in the lives of a young people's group, such as the arrival of new members.

Celebrations may not all be wild and exciting. Some will be solemn or leave us full of tears. And ways of marking these occasions will not all involve party-poppers and loud music. This resource aims to help you celebrate each event in ways which are appropriate to the occasion and to the young people.

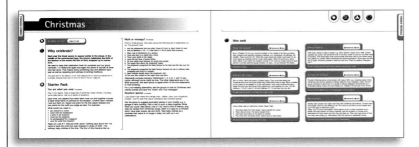

Most of all, *DIY Celebrations* seeks to ensure that young people see God's part in every event and respond to him. In the run-up to Christmas, it's possible that we read the Gospel accounts of the birth of Jesus with excitement and anticipation… but to be honest that's not really likely! We and the young people we work with are more likely to get bogged down with the trimmings and over-activity. And when did we last stop and thank God for the chance of a holiday? It is so easy to exclude God from our celebrations. This resource aims to bring it all back to him.

It's all very wobbly. Er, I think we mean flexible. There are over 200 ideas in *DIY Celebrations* which can be slotted into your programme in any way you want. Not to mention music tracks and clip art on the CD. That's why it's DIY. But if you want some Bible-based teaching programmes for the events through the year, we've provided *Starter packs* for nineteen such occasions to get you going. Along with each one there is an *Idea sack* with six brainwaves to adapt and use. Some are simple activities which would take up five minutes in a session. Others are full-blown celebration events in their own right. All aim to be adaptable to many different settings and numbers of young people.

In addition, at the back of the resource is a *Party bag* of additional ideas. Some are extras for the big occasions like Easter and Christmas. Others mark different events such as Bible Sunday or a visitor coming to the church. The contents list (page 5) and index (page 63) will guide you around.

What is included

What celebrations are included?

The celebrations covered in this resource are divided into occasions in the church year and events in the group members' own lives.

Church festivals

The round of church festivals helps us to focus on the different aspects of what God has done. It leads us to remember, re-explore and get excited about all the great parts of God's plan – from creation (remembered at Harvest) to Jesus' return (remembered in Advent).

The big, every-year events, like Christmas, Easter, Harvest and Pentecost, can challenge even the most creative of leaders with their regular reappearance. It's good to have fresh ideas for these each year. But perhaps we also need to invent or rediscover some traditions – things done in the same way every year which embody the truth of what God has done and which enable us to respond to him. Young people can relish these fixed points, provided they are done in a meaningful way.

There are other church celebrations, such as Epiphany, Ascension or Advent, that are easily forgotten or ignored in young people's groups. By marking these in engaging ways we can open group members' eyes to more of what God has done. If we always ignore these events, we may sell young people short. For example, Jesus' return is something to be looked forward to, celebrated and responded to. Just because it raises hard questions doesn't mean we should ignore it.

We have selected eleven church-year events. Ideas for others, such as Trinity Sunday and Bible Sunday, can be found in the *Party bag*.

Days in the life

In every group every year there will be young people who experience happy events and others who face difficult challenges. The resource material covering these events is designed to show God's involvement and care in every circumstance of life. That care will be seen lived out in the concern of group leaders and other people in the church. And it can be reflected in the support of group members for each other – sharing joys and struggles. There are activities suggested to encourage this mutual support.

Marking the regular events each year in the group – birthdays, start of the school year, and so on – is a powerful way of showing that God isn't simply interested in what we do at church. He's just as concerned about how we use our holidays.

There are also special times in the life of the group – the arrival of new members, a group 'official' birthday or a distinguished visitor. An event in the life of a leader – such as a new baby or house move – can also have a special feel for the group as a whole. Letting the group share in our own celebrations will build relationships and show how we make God part of those events (or it will at least encourage us to do so!). As the group members look to the leaders for guidance and example, they will notice how the leaders cope with the good and bad things that life brings along.

Please don't be daunted by the thought that you must celebrate all this at once. No one can do it all. View it instead as a range of opportunities each of which can open your group members' eyes to new things about our amazing God.

Contents of the book

Audio tracks

1.	Seven Days	3.29	Julia Kelly	Advent
2.	Mighty Ruler	4.36	Dave Woodman	Advent/Christmas
3.	Everything I Do	3.08	One	New Year/Start of school year
4.	The Prodigal	4.10	One	Lent/Tests and exams
5.	A Prize for Losing	3.28	Julia Kelly	Lent
6.	Forever	4.42	Richard Cimino	Easter Sunday
7.	Wake Up	3.00	One	Pentecost
8.	Thank You for the Rain	3.34	Justin Thomas	Tests and exams
9.	I'll be there	4.04	The World Wide Message Tribe	Change of school
10.	Never	3.03	One	Moving away

Electronic text

The full text of *DIY Celebrations*.

• To access, follow the instructions on this page and the next.

Images

• DIY Celebrations clipart images
• DIY Celebrations text icon buttons
• DIY Celebrations visuals

  Church festivals visuals for Advent, Christmas, Epiphany, Lent, Good Friday, Easter Sunday, Ascension, Pentecost, Harvest, Remembrance, All Saints

  Days in the life visuals for New Year, Birth and birthdays, Start of the school year, Tests and exams, Change of school, Holidays, Moving away, New group members

• 2000 clipart images and photographs from Greenstreet

Acknowledgements

TRACK 2 Dave Woodman ©1999 Daybreak Music Ltd, Silverdale Road, Eastbourne BN20 7AB.
Original recording from the album 'Resource 2000'; used by kind permission of Movation Records and Alliance Music

TRACK 6 Richard Cimino ©1996 Chief Musician Publishing.
Original recording from the album 'Resource 2000'; used by kind permission of Movation Records and Alliance Music

TRACK 8 Pennells/Porter/Thomas ©1998 Perfect Music UK/Alliance Media.
Original recording from the album 'Jumping In The House Of God Volume 3'; used by kind permission of Movation Records and Alliance Music

TRACK 9 Danté/Wanstall ©1999 Perfect Music UK/Alliance Media.
Original recording from the album 'Frantik' by The World Wide Message Tribe; used by kind permission of Movation Records and Alliance Music

TRACK 1,5 ©1999 Julia Kelly | **TRACK 3,4** ©2000 A Quinn | **TRACK 7** ©2000 R Magee/A Quinn | **TRACK 10** ©2000 A Quinn/D Thompson | **TRACK 3,4,7,10** recorded at Wildgoose Productions. Website: www.wildgoose.uk.com

Reading the electronic text

The CD text is in the same format as used on the internet and may be viewed using a web browser such as Microsoft's Internet Explorer or Netscape's Communicator. You do not need to be connected to the internet while you are browsing the CD. Because the text uses a standard web format, you may use almost any computer that has a recent web browser installed. (If you do not have a web browser on your computer, you may install one from the CD: read the file help.txt on the CD into any text editor or word processor and follow the instructions.)

• On Windows 95, 98 or later Windows computers, the CD should automatically start when you place it in the computer's CD drive. A menu will appear which will allow you to view the text in your web browser and install the software for viewing the images.

CD installation

- If the CD does not automatically start on your Windows 95/98 computer, run the menu software by clicking Start on your desktop. Select Run and type in x:\diy.exe, where x is the drive letter of your CD drive. (On many computers the CD drive is d, so you would type d:\diy.exe.) Click on OK. A menu will appear which will allow you to view the text in your web browser and install the software for viewing the images.
- If you are using a Microsoft Windows 3.1 computer, open the Program Manager and click on File and then Run. Type x:\win31.exe, where x is the drive letter of your CD drive. (On many computers the CD drive is d, so you would type d:\diy.exe.) Click on OK. This will run a web browser from the CD which will allow you to read the text. Please note that the image software provided on the CD will only work on Windows 95/98.
- If you are not using a Windows-based computer, check your computer can read CDs that use a Windows format. Most recent computers can. You should also use an up-to-date web browser that is capable of displaying tables and frames. Start your web browser and open the file index.htm on the CD.

The text on the CD is designed as a small web site and works in the same way. To move from section to section, click once on the buttons in the navigation pane on the left of the screen. Wherever you see text underlined, you may place your mouse cursor over the text and click once to move to that page or part of a page. To move back to a page you viewed earlier, click on the back button on the toolbar at the top of the screen. In places the text may differ from, or be more extensive than, the printed version.

Listening to the music

There are audio tracks which may be played through a standard audio CD player. You do not need a computer to play them.

Viewing the images

The images are stored in the greenst folder on the CD and may be viewed with most image programs. The clipart images are in WMF format and the photographs are in JPEG format. For those using Microsoft Windows 95 or later, Greenstreet have provided an image browser and there is some free image software in the software folder on the CD. This software may be installed from the opening CD menu.

Copyright and security

The aim in providing an electronic text is to encourage its use and adaptation for particular groups. However, all material, including software, is provided for the use of the purchaser of *DIY Celebrations* alone for use with his/her church and church groups. In copying text to a word processor, please respect this copyright restriction and also acknowledge on printed texts the source: for example, "From *DIY Celebrations*, CPAS". CPAS acknowledges all copyrights and trademarks, and material placed on this CD is either copyright CPAS or is used by permission.

The images on the CD are provided by Greenstreet and are copyrighted by them. They may be individually used in your own publications (paper or electronic), but may not be distributed as a collection.

CPAS cannot accept any liability for the software on this CD nor provide support in installing or running a web browser or the additional programs placed on the CD. The CD is checked for viruses during its production, but you are advised to use your own virus checker. CPAS cannot accept any liability for damage caused through viruses.

If you have problems...

If you have problems starting the CD, contact CPAS Distribution on 01926 458400 or e-mail sales@cpas.org.uk.

CPAS cannot support the free software supplied on the CD. You could approach the original supplier, but please remember that the software has been supplied to you free of charge and suppliers may not be able to provide support.

Celebration year planner

Use a copy of this planner each year to get an overview of the main events you plan to celebrate and the others you want to remember too, such as group members' birthdays.

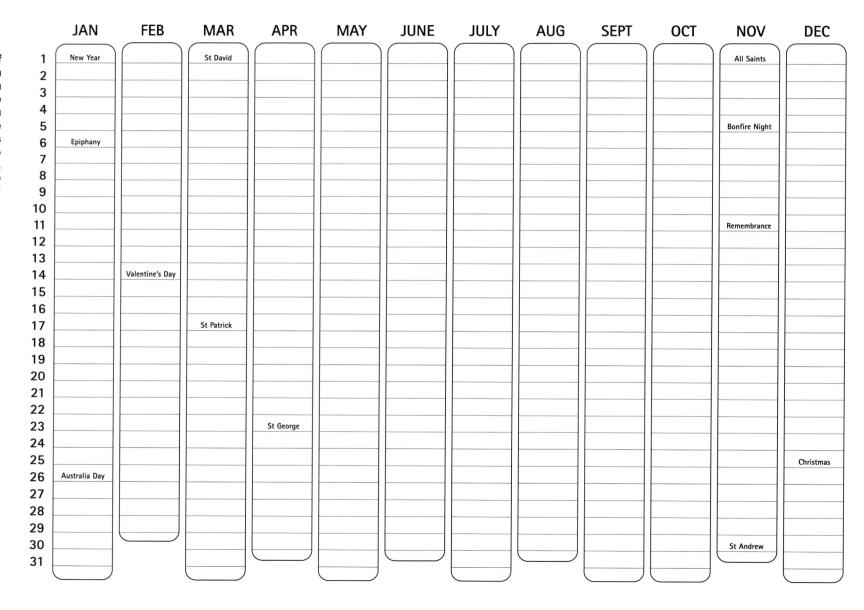

	JAN	FEB	MAR	APR	MAY	JUNE	JULY	AUG	SEPT	OCT	NOV	DEC
1	New Year		St David								All Saints	
2												
3												
4												
5											Bonfire Night	
6	Epiphany											
7												
8												
9												
10												
11											Remembrance	
12												
13												
14		Valentine's Day										
15												
16												
17			St Patrick									
18												
19												
20												
21												
22												
23				St George								
24												
25												Christmas
26	Australia Day											
27												
28												
29												
30											St Andrew	
31												

For each of the nineteen main celebration occasions in *DIY Celebrations* there are the following elements:

Key Bible passage
This underpins both the reason for celebrating and the learning and response activities for the celebration. Do some work on it first.

Why celebrate?
It's important to think clearly and pray about why we are celebrating, and to let this affect the way we celebrate so that it makes the event meaningful and memorable.

Starter pack
This consists of three ideas based on the Key Bible passage: introducing it, learning from it and responding to its truth. You can use any idea on its own, but they are also designed so that they can be used together for a brief but balanced learning slot.

Idea sack
These are a mix of ideas linked to the celebration. Some would fit into a learning-style of session, others are complete stand-alone celebration events, while others help the celebration to spill over to others – in the church through services or extra activities, or beyond the church to friends and families of group members. Each idea has preparation guidance. A tick indicates 'prep time necessary'; a cross indicates an 'instant' activity. Similarly, a cross by 'BOOK' suggests that the activity is appropriate for all, regardless of reading ability.

The *Party bag* at the back has ideas for extra occasions and some additional ideas for the nineteen already covered. Use the index to track these down.

Pick an occasion

Be selective in what occasions you celebrate. Plan the main events a year at a time so that you can spread them through the year in the way most helpful to you. Look back and decide which celebrations you want to repeat – inventing your own group 'traditions' about how you celebrate them. But also plan some one-off celebrations. Why not meet specially for Ascension Day (or whatever) just this year?

Alternatively, as a wacky one-off, you could blitz the whole church year in nine consecutive sessions – from Advent to All Saints. Go all-out to give each session a 'seasonal' feel even if it is the wrong time of year.

Menu options

Here are some suggestions about how to use the ideas:

Complete 'learning-style' session. You could use all the *Starter pack* ideas in the order given (or adapt or replace these with other activities – only use what will be best for your group). Mix in other items from the *Idea sack* and your own brainwaves.

Stand-alone celebration event. Use an *Idea sack* item which is a complete event or add together several ideas to make up your own event. Add in one or two thought-provoking activities on the meaning of the celebration – from the *Starter pack* – or perhaps a sharp 'epilogue' – such as reading the *Key Bible passage* and closing with prayer.

Quick cell. Pick just one short celebratory item to add into a group session programme when you have an occasion to mark but not long to do it.

Bible slot. Pick one or more *Starter pack* ideas for a short Bible slot in a seasonal session for a specific purpose – e.g. a rehearsal for a Christmas performance.

Church service. Many of the ideas can be adapted for use with a church congregation on a festival occasion. The group members could plan, lead or assist with this having first tried the activity themselves.

Celebrating

Celebrating with young people

Young people enjoy celebrating. And celebrations are big in the Bible. So it seems we're starting off from a great point of contact. Let's think about young people first.

It's worth asking any young people you work with how they would choose to celebrate something wonderful happening in their personal life. What would they do? Who would they share it with? Their ways of celebrating can be very different from those of their parents or of younger brothers or sisters. In celebrating with young people, we need to let them do it in their own way so it is meaningful to them.

With all this celebratory energy and vitality around (if that's the way to describe a sleepover!) it's right that we should make use of it by challenging the young people to celebrate everything that God has given them. That may mean being stunned by the wonder of Christ's birth in a totally new way, and it should also include allowing God to come into all the 'ordinary' events of life – the ups and downs which we celebrate and mark in different ways. For example, a visit to a favourite burger bar might follow both winning a sports trophy or a hospital appointment.

So what do young people have which we should take into account in thinking about celebrating with them?

- **Energy and vision.** Young people are able to devote time and gain knowledge in a way which is not possible as they get older. They can get excited and enthusiastic about any project they believe is right, and they can do a great deal of good.

- **Time to make decisions.** As they look to the future, young people have options of exam choices and careers to think through, and they have an opportunity to think about where they fit in the world now and where they hope to in the future.

- **Time to ask questions.** They may want to ask why we go through the motions each year and season, or what it all really means. They may question their faith and values, and challenge us to come up with some real answers. They often come through this with a stronger, more determined foundation for the future.

- **Relationships as a priority.** Young people find comfort and support in the challenges they face in life if they are surrounded by peers. They enjoy group activities, and can be good at drawing others into their groups. And, most of all, they feel able to relax and let their hair down in the non-judgemental environment they create for themselves.

- **Real lives.** It's not all rosy. Young people can have big issues about self-image and sexual development to struggle with. There may be painful memories of abuse, pain and hurt from the past. Some young people may feel responsible for family break-up and failure. If we are going to rejoice with them, we also need to mourn with them (Romans 12:15) – though this will be seen in careful pastoral attention rather than in a group setting.

Celebrations play a big role in the Old Testament. Take the Passover, for example, in which God's people were to remember how he rescued them from Egypt. Detailed instructions were given for the meal (e.g. Exodus 12:1-28). It wasn't just to become a big party. The whole thing was designed to make sure that everyone remembered what God had done. It begged the question 'Why are we doing this?' (Exodus 12:26) and its corresponding answer: 'Because the Lord...'

Later the celebration of the Passover was a marker of the Israelites' relationship with God. It was celebrated when they first entered the Promised Land (Joshua 5:10-12), its rediscovery was a sign of coming back to God in the time of King Josiah (2 Kings 23:21,23) and took on special meaning for those restored to the land after the exile in Babylon (Ezra 6:19-22).

The Old Testament celebrations were meant to form the framework of life, as God was at the heart of it. They were all laid down by God's command. But under the new covenant, Jesus gives instructions about only two celebrations – the commemoration of his death in the Lord's Supper or communion, and the marking of new life in him by baptism. That does not mean that all other celebrations are wrong. But it does mean they aren't compulsory. So why bother with them? (Check out the verses shown to think further about these.)

Celebrations are powerful reminders. We are much more likely to remember something we do than something we hear. If we hear about Jesus' return it has an impact at that time, but then vanishes from our minds at alarming speed. If we do something related to his return, we are more likely to remember it. If that action is packed with excitement or wonder, and if we repeat it, it can become part of our whole framework of life (Deuteronomy 4:9-10).

Celebrations are ways to respond. What God has done is amazing. So is what he is doing in young people's lives and what he will do when Jesus returns. It all demands a response from us. Or rather, lots of different responses – joy, enthusiasm, sadness, action, speaking, silence... (Psalm 145:1-3).

Celebrations communicate. They have an impact on those who are involved and those who are aware of them from outside. So they are great for communicating about God to young people – passing the good news on from one generation to the next – and from young people to their friends, families and beyond. They are a great excuse for inviting friends along and saying something about the point of the celebration. Celebration can also 'infect' the rest of the church, and young people can help to plan how to make this happen (Psalm 145:4-7).

Celebrations value individuals. Most of us really enjoy being invited to a party or celebration. The fact that we are invited makes us feel valued and part of something bigger. We can also celebrate individual young people at birthdays or when they arrive or leave the group. In those celebrations we reflect God's loving concern for individuals (Romans 12:9-16).

Celebrations build relationships. Any group of young people will 'gel' more by having fun together. But celebrations focused on God will lead to much deeper relationships built on the great things they share together – all that God has done for them and for us. By celebrating each other's joys and supporting each other through tough experiences, they and we will be showing God's care and compassion (Ephesians 4:1-6).

Celebration is good. Very good. It focuses energy and vision on God, it motivates and draws together, it brings fun and enjoyment into lives, it raises questions and gives answers, and points us and others to the endless good things that God gives. So go on – celebrate!

Visuals

Making it visual

From here to page 17, you'll find a selection of images to accompany the nineteen topics included in the Church Festivals and Days in the Life sections of *DIY Celebrations*.

For good measure, on this page we've also included clipart images of *DIY*'s text illustrations and 'icon buttons'.

Feel free to photocopy any of these images for use in publicity leaflets, programmes, posters, OHP transparencies, whatever. It's up to you. And don't forget that all of these visuals are available in electronic form on the CD. See pages 6 and 7 for full instructions.

Clip art

Advent

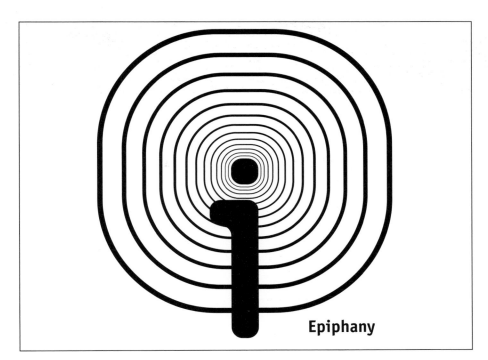

Epiphany

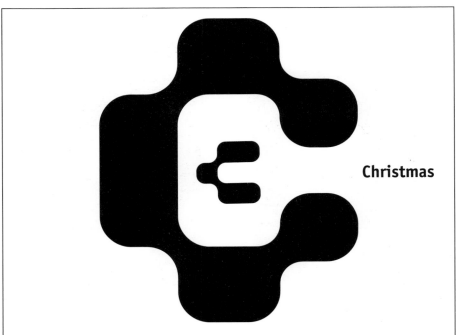

Christmas

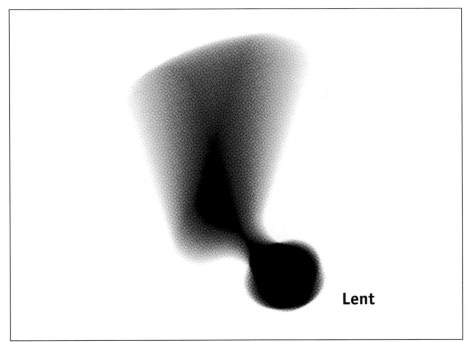

Lent

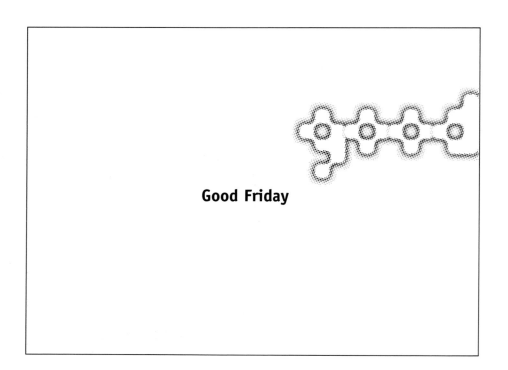

Good Friday

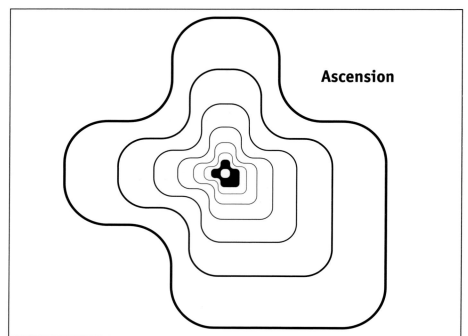

Ascension

Easter Sunday

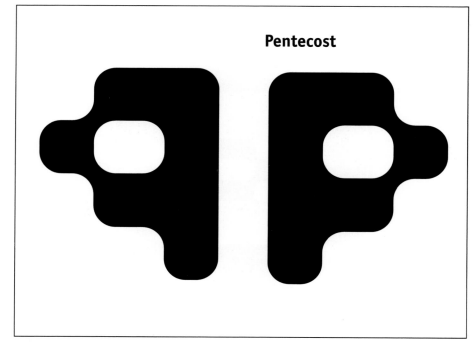

Pentecost

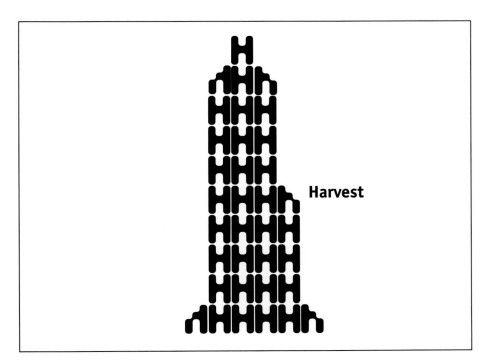

Harvest

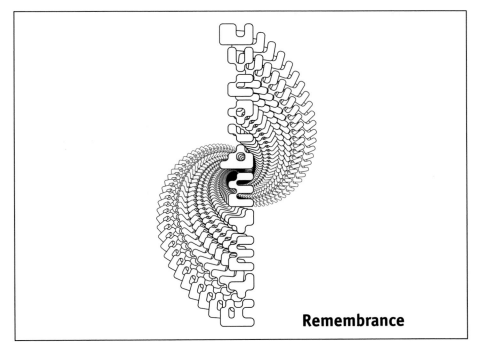

Remembrance

All Saints

New Year

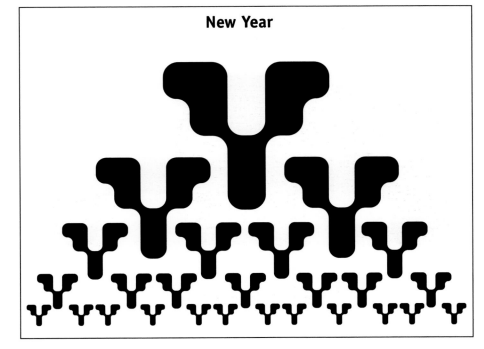

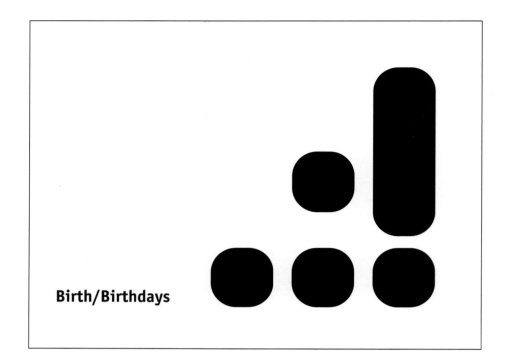

Birth/Birthdays

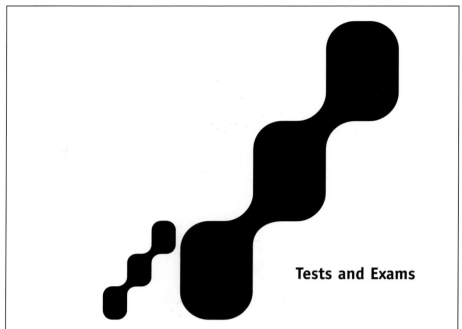

Tests and Exams

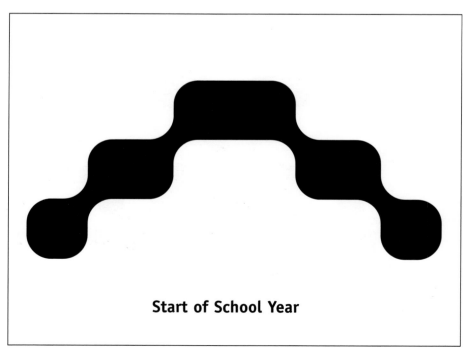

Start of School Year

Change of School

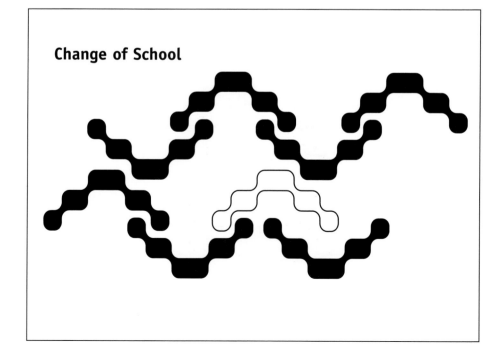

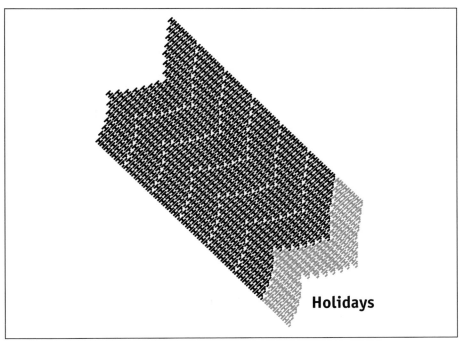

Holidays

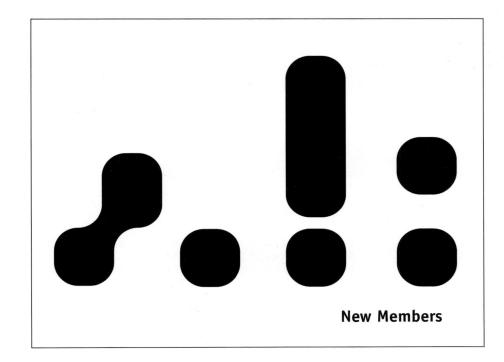

New Members

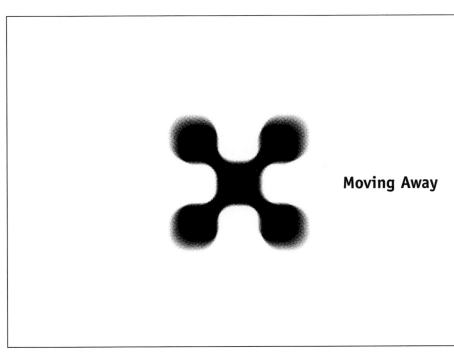

Moving Away

Advent

KEY BIBLE PASSAGE *Isaiah 9:2-7*

Why celebrate?

Advent is the time of preparation for Christmas, starting four Sundays before 25 December. There can be so much to do at this time of year. But Advent should be a time to pause and take stock, to take a fresh look at the coming of Jesus into the world and to prepare ourselves for the time when he will come back. We need to make space to do this.

Isaiah spoke of a great light coming to those who live in darkness. Someone was on his way who would reign for ever, bringing justice and peace. All of us, including our young people, need to feel the joy he brings.

'The people walking in darkness have seen a great light… You have… increased their joy.' Isaiah 9:2–3

Starter pack

The Preparation Game (15 minutes)

Split into teams of two to four people. Have the following items for each team: loaf of bread, spreading knife, jar of jam, large box, roll of Sellotape, Christmas wrapping paper, large piece of card, marker pen.

Give the teams seven minutes to complete all the following tasks in any order:

Task 1 Make as many jam sandwiches as possible, collecting one slice of bread at a time. (Position the loaf outside or in another room if possible.)

Task 2 Gift-wrap the large box attractively.

Task 3 Use the card and pen to make a Christmas card which says something about the real meaning of Christmas.

Task 4 Think about what it might be like when Jesus comes back.

At the end discuss the results. Did any tasks get missed? Say a little about what Advent is and use the following discussion questions:

- How can we as a group prepare to make this Christmas a real celebration of Jesus' birth?
- What can we do as a group this Advent to help us be ready for when Jesus comes back?

Prepare the Way (15 minutes)

To prepare, make photocopies of Isaiah 9:2-3 and Isaiah 9:6-7 for each pair. Cut up each section into individual lines, removing the verse numbers, and put the pieces in two separate envelopes.

Read out Isaiah 9:2–3. Give each pair their first envelope and ask them to arrange the pieces in the right order.

Read out the correct version again. Ask the group members to suggest what these verses might be about – what sort of darkness do people live in, and how does God bring light?

Read Isaiah 9:6-7 and repeat the activity with the second set of envelopes.

Explain that these verses say how the Light would come. Give each pair two minutes to decide 'Is it all true?' then discuss their ideas. Explain simply how Jesus fulfils the prophecy, bringing light to us. For example:

- 'Wonderful Counsellor' – he teaches us the truth
- 'Prince of Peace' – he puts things right
- 'His royal power' – he's in charge of everything. One day everyone will see that.

My Preparation (10 minutes)

Ask each group member to think of ten important things to do before Christmas. They could write them down. Encourage them to share ideas with those near them.

Have a pile of small candles. Light one and ask the young people to sit quietly and think about being ready for when Jesus comes back. Read out Isaiah 9:2. Then invite each young person to come forward and take a candle to symbolize their personal commitment to remember Jesus during Advent.

Idea sack

Flickering
PREPARATION ☒ BOOK

All sit around a lighted candle. Ask the group members to say as many words as possible which describe the candle or flame. Remind them that Jesus described himself as the light of the world. Still looking at the flame, ask them to say words which describe Jesus.

To think about what it means that Jesus is the light of the world

Advent lights
PREPARATION ☒ BOOK

Buy an Advent candle. Each week in Advent, darken the room and light the candle. Encourage the young people to close their eyes or look at the candle while you read from the Bible. Use a different passage each week: John 1:1-14; 1 Thessalonians 4:15-17; Revelation 21:1-4; Luke 2:1-7. Some suitable quiet music may help them focus on Jesus.

To prepare for Jesus at Advent

Coming again
PREPARATION BOOK

Assign small groups Bible passages relating to Jesus' return: Matthew 24:36-44; 1 Thessalonians 4:13 – 5:4; Revelation 20:11 – 21:4. Ask them to find out what they can about Jesus' return using the following questions:
• What might it be like?
• Is it possible to predict when and where it will happen?
Also ask the groups to come up with their own questions about Jesus' return. Invite a guest, e.g. your church's minister, to answer their questions in a clear and reassuring way.

To think through Jesus' return

Prepare to breakfast
PREPARATION ☒ BOOK

Check with the group members what they most like for breakfast – get a few imaginative ideas as well as their usual ones. Arrange to meet for breakfast early one morning. Provide the ingredients yourself, but get the group members to prepare and cook them. As they finish eating, read Luke 14:15-24. Explain that Jesus' return will be a bit like a great feast – he's preparing it and everyone's invited. Pray for people you know who have not yet said 'yes' to his invitation.

To look forward to Jesus' return

Candle factory
PREPARATION ☒ BOOK

Run a candle-making workshop. Either get someone who's done it before to help, or get hold of a book on candle-making (see Resource page 00). Make some candles for use in church at Christmas. Group members may want to make some as Christmas presents, too.

To reinforce learning about Jesus as the light

If I could give
☒ PREPARATION BOOK

Ask the group members to think about gifts that they are planning to give to friends and family at Christmas. Give each one as many slips of paper as there are people in the group. Ask them to write on these the gifts they would give each of the others. They should all then exchange 'gifts'.

Ask: How do you feel about the gifts you got and those you gave? Challenge the group members each to fill in the last slip privately with what they would like to give Jesus this Christmas – what they would like to be or do for him. Some might want to share what they have written.

To see Christmas presents in perspective

Christmas

Why celebrate?

Each year the tinsel seems to appear earlier in the shops. In the middle of the commercial hype, the church celebrates the birth of the Saviour of the world: the Son of God, wrapped up in human form.

We need to keep that celebration fresh for ourselves and our group members – to rediscover again and again the sense of wonder at what God has done. That means doing something new and different each year as well as repeating (and perhaps inventing) traditions.

'She gave birth to her firstborn, a son. She wrapped him in cloths and placed him in a manger, because there was no room for them in the inn.' **Luke 2:7**

Starter Pack

You are what you wear (15 minutes)

Play in two teams. Have a large pile of extremely varied clothes, including some baby items. Call out a series of situations.

Each time, two players from each team must run and together choose at least three items of clothing for the situation. Another team member must put them on. Award one point for the first player dressed and two points for the outfit you judge as best. For example:

What would you wear to…

- get noticed in a crowd
- lunch with the headteacher
- a party in the Antarctic
- muck out the elephants
- be photographed for a magazine
- save the world from disaster?

Read out Luke 2:7. Discuss what Jesus' clothing says about him. He came to save the world but was wrapped in strips of cloth – the ordinary baby clothes of the time. The Son of God became like us.

Myth or message? (15 minutes)

Work in small groups. Give each group the following set of statements cut up. The groups must:

- sort the statements into two piles: 'know it's true' or 'don't think it's true'
- look at Matthew 1:18 – 2:1 and Luke 2:1-20 to check their answers.

1. Mary rode to Bethlehem on a donkey.
2. Jesus was born in a stable like a shed.
3. There were cattle in the stable.
4. Jesus did not have a human father.
5. He was called Jesus because he would save people.
6. The angels flew above the shepherds' heads.
7. The shepherds recognized the baby Saviour by his halo and the star over the roof.
8. The shepherds recognized the baby Saviour because he was in ordinary baby wrappings and lying in a manger.
9. Mary thought deeply about the shepherds' visit.
10. The wise men visited on the night Jesus was born.

Afterwards compare answers. Statements 1, 2, 3, 6, 7, and 10 are myths, although some *might* be true. The other statements are in the Bible accounts. Ask each group to agree which of the true statements is most amazing.

For a non-reading alternative, ask the groups to look at Christmas card nativity scenes and spot the 'myths' and 'true messages'.

Nowhere special (15 minutes)

Luke doesn't say where the manger was – stable, cave, poor household, outside – but it's clear that Jesus' birthplace was nowhere special.

Ask the group to suggest equivalent places in your locality, e.g. a garage or farm building. Plan a visit to such a place together. While there you could: read aloud Luke 2:1-20, have a time of silence, sing, read out verses from Christmas songs, listen to a tape by a Christian band, have open prayer, eat and drink. Take baby clothes as a reminder that Jesus is no longer a baby, but with us in our celebrations.

Idea sack

Pack the parcel!

☒ PREPARATION ☒ BOOK

Buy a Christian CD for your church's minister. In the middle of the floor put sticky tape, scissors and assorted wrapping materials, e.g. newspaper, old giftwrap, foil, clingfilm. Pass round the CD. When the music stops the holder must wrap it in as many layers as possible until the music starts again. When it's enormous, add a label, deliver it and watch the recipient unwrap it. For more fun make everyone wear gloves or provide assorted gunge (custard, oil, glue...) to put between the layers.

To enjoy giving

Tell me the story

☒ PREPARATION ☑ BOOK

Sit in a circle. Hand one person a (clean) nappy. They must start telling the Christmas story from the point where Mary heard she'd have a baby. Blow a whistle for the nappy to be passed on to the next person, who must continue the story. Make the turns very short (even half a sentence). Look together at Luke 1:26-38, Matthew 1:18-25, Luke 2:1-20 and Matthew 2:1-12 and see how you did.

To get into the Bible's account of Jesus' birth

Carol quiz

☑ PREPARATION ☒ BOOK

Have a team quiz on well-known carols. Keep it light.

1. Give lines taken from first verses. Teams identify the carols.
2. Read obscure lines. The teams explain them.
3. Give lines backwards for the teams to identify.
4. Play intros from recorded carols. Teams race to name them.

To think about the meanings of carols

Virtual babies

☑ PREPARATION ☒ BOOK

Write baby needs on slips of paper, e.g. drink, blanket, nappy, food, rattle, jumper, squeaky toy, socks. Ask someone to lie on the floor under a duvet so only their head is showing. This virtual baby must communicate what they need to the others using only yelling (no words), head movements and facial expressions. Show them the first slip of paper. Whoever guesses it does the next one. Finish by reading Philippians 2:6-7.

To get inside what it meant for Jesus becoming a baby

A time for Christ?

☑ PREPARATION ☒ BOOK

Provide Christmas items unconnected with Jesus' birth, e.g. wrapping paper, tinsel, Christmas tree branch, pictures of Santa, balloons, food wrappers, cards. Get the group members to stick them on a length of strong wallpaper. Somewhere put a small label saying 'Guess who got missed out.' Display it for the rest of the church.

To challenge people not to forget Jesus at Christmas

Decorations

☑ PREPARATION ☒ BOOK

Identify older people who might want help with Christmas decorations. Contact them and explain what you are offering to do. Fix a time to put them up and explain that you'll visit again to take them down.

Make, buy or borrow decorations. Send teams of two or three members and a leader with stepladders and fixing materials. Ask the recipient to choose what decorations to have where and how they are fixed up. They may have some of their own they want putting up. Alternatively offer this service to residential homes.

To encourage older members of the community

Epiphany

KEY BIBLE PASSAGE   *Matthew 2:1-12*

Why celebrate?

We easily miss the true impact of the visit of the Magi, or 'wise men', to Bethlehem. The events get absorbed into Christmas activities. And by 6 January, the traditional date for marking this event, most of us have had enough of it all.

But it is stunning the way God revealed the birth of his king to these unlikely star-experts in a distant land. It shows that, in this baby, God was revealing himself to the whole world. Everyone in every part of the world should kneel before Jesus. These events should inspire us to praise God for the worldwide spread of the good news, and challenge us to be part of ensuring that others see and hear what he has done.

'We saw his star in the east and have come to worship him.' **Matthew 2:2**

Starter pack

Go or no? (10 minutes)

Label the ends of the room 'go' and 'no'. Call out the following situations. The group members must decide whether they would go and run to the appropriate label. Each time, ask one group member to explain their decision. Make up further situations.

- A stranger offers you a lift home from school.
- A friend tells you that a local shop is giving away chocolate bars.
- A mate invites you to go hang-gliding.
- You're lost in a wood. You see a small side path you haven't tried yet.
- You are sent tickets to the biggest gig ever – all the best bands. But it's 400 miles away.
- A time travel machine is being tested. You can go back and visit Jesus as a baby, but you may not be able to return.

After the last of these, ask several group members to explain their choice. Go straight into reading Matthew 2:1-12.

Surprise visit (15 minutes)

In threes or fours, read Matthew 2:1-12 and discuss the following questions. Give each group a stone to pass round as they discuss. Only the person holding the stone may speak.

- Why did the visitors come to find Jesus?
- Why did Herod panic?
- What had the prophet Micah said about the Messiah?
- Why was Herod so keen to know where the new king was?
- Why did God warn the visitors to take another route home?
- Choose your 'most surprising fact' from all that happened.

Take quick feedback from the groups. Use the last question to highlight the 'surprise' that God revealed the birth to these unlikely foreigners and how they responded.

For all the world to see (15 minutes)

Celebrate the worldwide spread of the good news of Jesus by using the Internet and e-mail. Ask for help with the technical side if necessary.

If the group is small, set up a computer to use in the session. Gather round and take turns at the controls. Alternatively, ask interested group members to do the computer tasks at home and to report back in the session.

The tasks are:

- List all the countries you can show that have Christians in them, in five minutes' surfing.
- Find out how people in one other country have celebrated Jesus' birth this Christmas.
- Ask Christians in other countries: 'Why do you celebrate Jesus being born?'

Pray using the information gathered. Name countries, people or topics and allow a brief time of silence after each. Pray for those who have not yet heard about Jesus.

Idea sack

Epiphiparty
☑ PREPARATION ☒ BOOK

Save your Christmas celebration for around Epiphany:
- Have international food, music, etc.
- Put up fluorescent stars from a stationer's.
- Do a treasure hunt. Position adults wearing stars round the area, each holding an object as a clue to the next place.
- By candlelight read Matthew 2:1-12.
- Give gifts with a 'gold' connection.

To celebrate Jesus' birth in a different way

News view
☒ PREPARATION ☑ BOOK

Ask the group to watch the news during the week before the session. Give pairs or threes (or confident individuals) different parts of the world to report back on as foreign correspondents (e.g. the different continents). At the start of the session ask the groups of correspondents to stand in front of an imaginary camera and file a quick report. Later in the session pray for those countries.

To hear and pray about the world to which Jesus came

Godwords
☒ PREPARATION ☒ BOOK

Encourage the group to come up with as many words as possible to describe God – lots, not just the obvious. Write their ideas on a wall or fill a page of the church magazine with a scatter of handwritten words. Each time they reach a 'block', change to one of the following activities, then get going again:
- eat something sweet and describe the experience
- ask what God has done for them
- blow up balloons till they burst – describe the experience.

To express what they know of God and reveal it to others

Palace Bethlehem home
☑ PREPARATION ☒ BOOK

Do a Magi meal trip and sleepover. Starting from your usual group base, members visit three different venues in turn, carrying their sleeping bags and nightgear:
palace – a big(gish) home with lavish Christmas decorations, loud music and great savoury food. Read Matthew 2:1-8.
Bethlehem – a simple room in a small home with reflective music, a baby(?) and simple dessert items such as fruit. Perhaps sing worshipful songs. Read Matthew 2:9-12.
home – your usual group base with cake and supper drinks, games and a sleepover.

To celebrate Jesus' birth

Prayer stars
☒ PREPARATION ☒ BOOK

Ask the group to lie on the floor in circles of up to six with heads in the middle – forming stars. Encourage them to pray in turn round the circle – just one word or one sentence each. Announce prayer topics: what I like best about Jesus (praise); people who've told us about Jesus (thanks); countries; people in our church who tell others about Jesus.

To pray aloud without embarrassment

World's a balloon
☒ PREPARATION ☒ BOOK

Challenge each group of three or four to turn a balloon into a globe by drawing on the continents with a marker pen. The knot is the North Pole. Ask them to put a sticky label on any part of the world where they know someone. Hold bright torches against the balloons at the South Pole. Darken the room and read verses about Jesus as light for the world: John 8:12; 1:5; 3:19; 2 Corinthians 4:6.

To see Jesus as the light of the real world

Lent

Why celebrate?

Lent lasts six and a half weeks from Ash Wednesday to Easter Saturday. Like Jesus' time of fasting and temptation in the desert, it's a special opportunity to face up to the things most likely to pull us away from his purposes. Each of Jesus' answers to Satan speaks of wholehearted commitment to his Father God's plan – and his thorough knowledge of Scripture. The point of Lent is to follow him – to reset the focus of our lives on God and his purpose for us through prayer, Bible reading, reflection, confession and knowing his forgiveness.

Fasting can help to show that we love God more than any of the wonderful things he's created (verse 4). But it is not an end in itself. Giving up chocolate probably won't transform the lives of our group members. Spending serious time with God will.

Jesus said to him, "Away from me, Satan! For it is written: 'Worship the Lord your God, and serve him only.'" **Matthew 4:10**

Starter pack

Confession costs (15 minutes)

Beforehand make cards each with a 'crime' and a 'fine' on, e.g.

	£
arguing	150
greed	150
insults	250
disobeying parents	200
boasting	200
envy	150
lying	100
hurting others	150
selfishness	250

Set a time limit, e.g. 10 minutes. Sit in a circle. Give each player £500 of Monopoly money and a 'crime card'. A leader (the 'judge') bangs a table. The group pass their crime cards to the right until the next bang of the table. The judge calls out a crime at random. The person who holds that card must pay the amount shown. If they cannot pay they have 30 seconds to confess to a true example of that crime. If they succeed, the judge pardons them and makes their money back up to £500. If not, they are out. The person with most money at the end wins.

Ask where the game gave a true picture of God's forgiveness and where it was false. Jesus paid a much greater price for our sin. His forgiveness is always certain.

Deflectors (10 minutes)

Ask a group member to mime a task, e.g. digging a hole, cooking on a barbecue, playing the trombone, getting to sleep. The rest of the group must try to deflect them from the task by giving bad advice or saying why they should give up. The 'actor' must answer each comment. After five or six attempted 'deflections' have a new 'actor' with a different 'task'.

At the end ask what Jesus came to do. Then read Matthew 4:1-11 to show how the devil tried to deflect him. In silence thank Jesus for keeping on with what he came to do for us.

Lent give-away (10 minutes)

Ask the group members to suggest reasons why people give things up for Lent, e.g. to give time for prayer, to impress God. Write these up. Ask which are bad reasons and take a vote on the best reason. In the discussion explain the main point of Lent (see 'Why celebrate?'). Giving something up may help if it's to show God that we love him more than the best things he's created (like chocolate).

Ask the young people to write on one side of a slip of paper something that God might want them to give up, e.g. a moody attitude. Pause to ask God's forgiveness in silence.

On the other side invite them to write something practical they could give up for Lent – if that will help them focus on God. Pray, thanking God for all he has given us, for total forgiveness and most of all for Jesus.

Idea sack

Pancake mix

☑ PREPARATION ☒ BOOK

Have a pancake party. Provide twenty toppings, sweet and savoury, e.g. lemon, orange, banana, apple, marmalade, jam, sugar, chocolate chips, tomato ketchup, ham, cucumber, tomato, pickle, crisps, honey, cheese, tuna, bacon, mayonnaise, yoghurt. Number these and hide them. Ask each person to draw two numbered slips of paper. Encourage but don't force them to try the two toppings together. (Watch out for any group members who suffer from food allergies – many products contain traces of nuts, for example.) Explain the origin of having pancakes – using up rich food before Lent.

To mark the start of Lent

Silence please

☑ PREPARATION ☒ BOOK

Have a half-hour silence to concentrate on Jesus at the start of your session or a separate time. Provide plenty for group members to do – ideas for prayer written up, an option of a silent walk round the area, the 'Jesus' video playing without sound, Bible reading notes, Bible references to look up (e.g. Matthew 4:1-11), a practical task to do. Afterwards ask how useful it was.

To focus on Jesus

Keeping on

☑ PREPARATION ☒ BOOK

Ask group members to tell what they know of the life stories of Jonah, Daniel, Simon Peter and Paul. Ask them what pressure there was for those people to give up on God's plans. Ask them to imagine that their life story so far is in the Bible. What would have been the main pressures on them to give up on God?

To be encouraged not to give up following God's purpose

Eat and drink

☑ PREPARATION ☑ BOOK

Split into pairs and give each pair a cup of water, a teaspoon, and a dry cream cracker. The pair must race, with one trying to drink the water with a teaspoon and the other trying to eat a dry cream cracker. Then ask the pairs to look up John 6:35 and John 4:14 and talk about what they mean for us.

To show the importance of Jesus in our lives

Action fraction

☑ PREPARATION ☑ BOOK

Through Lent, work on a brief dramatic production of Jesus' life, death and resurrection for Easter. Choose ten incidents from Luke. Use the passages from *The Message* translation as narration. Work on stylized action in which at any point most of the characters are frozen in an action snapshot pose. Just the one (or occasionally two) who is the focus of the action is moving – during the scene you can switch who this is. Have some incidents read by several readers with no action.

To explore and present Jesus' life, death and resurrection

Bubbles

☑ PREPARATION ☒ BOOK

Explain that washing is used as a picture of confession and God's forgiveness in the Bible. Read James 4:7-10 and Psalm 51:1-2. Give each person the choice of a tiny pot of bubble bath or shower gel (with a unisex smell or a choice). Suggest that some time in Lent they take a long bath or shower and use the opportunity for a spiritual clean-out, too: reviewing their lives, confessing and enjoying God's forgiveness.

To confess and know God's forgiveness

Good Friday

KEY BIBLE PASSAGE Mark 15:22-41

Why celebrate?

Mark's Gospel makes the humiliation and the anguish of Jesus' crucifixion very clear. But he also keeps reminding us of just who Jesus is (even through the words of the insults hurled at Jesus). He's the King of the Jews, the Saviour, the Messiah, the Son of God.

Our Good Friday celebration should have a sense of seriousness as we remember what Jesus went through for us. But it needs to acknowledge who Jesus is, and so it should be full of joy, too, because the risen Lord himself is with us and our young people, celebrating what his death has achieved. We are his people, forgiven, free. The way to God is open.

'With a loud cry, Jesus breathed his last. The curtain of the temple was torn in two from top to bottom.' **Mark 15:37-38**

Starter pack

Order (10 minutes)

Write the following events from Mark 14-15 on cards. You need one for each person. Select fewer or add more as necessary. Number the cards in the order shown.

Spread out the cards. Explain that when you say 'go' each person must pick one up. They have one minute to plan individually how to mime the event shown. Perform them in number order, saying what each one represents.

Repeat the activity, taking different cards.

- Jesus was arrested.
- The Council questioned Jesus. He said he was God's Son.
- Jesus was spat at, blindfolded and beaten.
- Peter denied knowing Jesus.
- Pilate questioned Jesus.

- The crowd shouted 'Crucify him.'
- Soldiers dressed Jesus up and mocked him.
- Simon of Cyrene helped carry the cross.
- Jesus was put on the cross at 9 a.m.
- People insulted Jesus on the cross.
- At noon it went dark.
- When Jesus died, the army officer realized he was from God.

This man was really... (10 minutes)

Write each of the following numbers on a separate card: 23, 26, 29, 32, 33, 34, 36, 37. Hand them out to pairs or small groups.

Read out Mark 15:39. Ask: 'What was it about the way Jesus died that made the army officer say this?' Get the group members to look up the verse numbers on the cards in Mark 15 for ideas.

Ask each group member to say in one word how they would have felt if they'd been a soldier standing near the officer. In one word, how do they personally feel about Jesus' death today?

Bad News or Good News (15 minutes)

Have two bags labelled 'Good News' and 'Bad News'. Ask the group members to tear items of good and bad news from a pile of recent newspapers, and place them in the correct bags. After five minutes, review what is in each bag. (Alternatively, show very brief video clips from the news and vote whether each is good or bad news.)

Ask the group members what makes news 'bad' or 'good'. Is the news that Jesus died on a cross good or bad? That event looks like the worst news ever, but actually it's the best news ever, because of what Jesus' death achieved for us – forgiveness, freedom, life for ever with God.

Read out Mark 15:33-39 with quiet background music. Spend three minutes in quiet reflection on what Jesus did for us.

Explain that just because it's Good Friday we don't have to forget that Jesus is alive. He's with us now, celebrating what he's done. Have a three-minute party with food, drink, music, party-poppers, the lot. Finish with a prayer thanking Jesus.

Idea sack

Drawing inspiration
☒ PREPARATION ☒ BOOK

Have a pad of small pieces of paper and pens/pencils available throughout the session or take them to a Good Friday service. Encourage the group members to draw what they want in response to what they hear in the Bible. If they are doing this in a service, check that the service leader is happy about this. Afterwards, some may want to share their drawings with others.

To encourage reflection on the events of Good Friday

Tense times
☑ PREPARATION ☑ BOOK

Have copies of 1 Kings 19:1-10 available for all the group. Read it out and think through the following questions in pairs or threes:
• Why was Elijah in trouble?
• How did he feel?
• Are there any similarities with how the disciples felt on Good Friday?
Come back together and ask: What can we do if we reach 'the end of the line'? Read Acts 16:16-34 for a very different response to trouble in the light of what Jesus did on Good Friday.

To explore the group's response to tense situations

Living witness
☑ PREPARATION ☒ BOOK

Encourage the young people to take part in a Good Friday church service. Liaise with the service leader about what would be helpful. For example, prepare poetry, drama or music on the theme of Jesus' death and what it achieved. Meet before or after the service for hot cross buns.

To reflect on Jesus' death with others in the church

Nailed
☑ PREPARATION ☒ BOOK

In the centre of the group place the following items: red cloth, thorns, a large stick, a cross made from two pieces of rough wood, three 13cm nails, a cup of wine/juice, dice, the label 'This is Jesus, the King of the Jews', a sponge on a stick, torn curtain material. Ask each group member to take an item to hold. They may exchange their objects for those in the centre at any point as you read Matthew 27:27-55. Encourage the group members to reflect on the love that made Jesus endure such shame and pain.

To think what Jesus endured for us

The flicks
☑ PREPARATION ☒ BOOK

Have two video nights. Before Good Friday show an enjoyable secular video – with a strong hero character. On or around Good Friday show the 'Jesus' video, perhaps starting at Jesus' baptism. You could provide popcorn and give it all a cinema feel. After the video, or next time you meet, ask what they thought of it. Chat about how Jesus compared with the hero of the other film.

To look at the life, death and resurrection of Jesus

Passover
☑ PREPARATION ☒ BOOK

Share a Passover meal together – looking back to Jesus' celebration of the Last Supper and the new meaning he gave the Passover celebration. Resources are available from Church's Ministry Among the Jews (CMJ), 30c Clarence Road, St Albans, Herts, AL1 4JJ. You could invite the members' families, if they would like this.

To experience a Passover meal together

Easter Sunday

KEY BIBLE PASSAGE | *Luke 24:1-12*

Why celebrate?

The women gazed at the empty tomb in confusion, followed by terror when two angels appeared. But the exciting message was that Jesus is alive. When the disciples heard the women's report they too were confused – unable to believe that Jesus had risen from the dead.

Some of our group members will be questioning and uncertain about Jesus' resurrection, wanting to know the evidence. That's normal and natural. But Easter is an essential part of our faith. If Jesus didn't really beat death then we have no hope in the world. Our celebration needs to help them experience the same joy of realization that the disciples felt. It's true, Jesus is alive, death has been defeated.

'…when they entered, they did not find the body of the Lord Jesus.' **Luke 24:3**

Starter pack

Eggs (15 minutes)

Give each group a sheet of paper and a pen. Explain that you will call out a word and they have two minutes to write down as many words as possible connected with it. You will then ask each group to call out the words on their list and cross out any that another group also had. Give each group a chocolate mini-egg for each word they are left with. Repeat the game with a different word. Try: 'eggs', 'tomb', 'alive', 'Jesus'.

At the end ask the groups to read again the words they had left for 'Jesus'. Point out that the disciples thought all of that was lost when Jesus died. If they'd been right then we'd have known nothing about Jesus. But they were wrong. All that Jesus is and all that he can do are for us – our Easter present – because his tomb was empty and he is alive today.

Seen again (15 minutes)

If you have enough people, split into four groups. Give each group one of the following Gospel resurrection accounts: Matthew 28:1-10, Luke 24:36-49, John 20:11-18, John 20:19-29. If you have fewer groups, leave out some of the Bible passages.

Give each group the following questions to answer:
- Who found out that Jesus is alive?
- How did they find out?
- How could they be sure?
- What did Jesus say?
- What did those people do at the time?
- What did they do afterwards?

Compare the findings of the groups. Point out that the evidence that Jesus is alive is very good. Ask the group members how they would answer someone who asked: 'Why do you believe all that stuff about Jesus? Once you're dead, you're dead.'

Photogravey (10 minutes)

Go to a location by your local graveyard (or somewhere else suitable) to do a photoshoot of key events in the story of Luke 24:1-12. Get the group members to do as much of the work as possible – plan what pictures to take, be the actors and handle the camera. Roughly one photo for each verse should be about right. Encourage them to make sure the powerful emotions of the story come across – misery, confusion, fear, shock, excitement, disbelief, amazement.

Plan to meet together as soon as you can get the pictures developed to create a storyboard with captions for the photos. Agree where to put it up so others can see it and enjoy the truth that Jesus is alive.

Idea sack

Tombstone graffiti

☒ PREPARATION ☑ BOOK

Put up a huge circle made from sheets of black sugar paper. Blu-Tack over it an identical circle of white paper. At the start of the session, invite the group to draw or write on it their thoughts and feelings about the death of Jesus. At the end move the white circle to the side. Ask them to use chalks on the black circle to record their thoughts and feelings about Jesus' resurrection.

To express feelings about Jesus' death and resurrection

Sunrise sniff

☑ PREPARATION ☒ BOOK

Meet up just before dawn on Easter Day (or have a sleepover the night before). Ask each group member to bring something nice-smelling. Walk to a local park or beauty spot. As the sun rises open the smelly items, reminding the group of the women who went to Jesus' tomb to put nice-smelling things on his dead body. Read Luke 24:1-7. Pray and sing before returning for breakfast.

To capture the excitement of Jesus' resurrection

What's the difference?

☑ PREPARATION ☒ BOOK

Read out Matthew 28:1-10. Hand out whistles, hooters, etc. Read out Luke 24:1-12. Each time a group member hears a difference from the first account they must make a noise and say what the difference is. Discuss the following:
- Do the differences make you doubt that Jesus did rise again?
- What reasons might there be for differences (e.g. perspectives of different witnesses)?
- Do they matter?

To grapple with the various resurrection accounts

RIP

☑ PREPARATION ☑ BOOK

Give each person a gravestone-shaped card with 'RIP' on it. Explain that these letters on gravestones stand for 'rest in peace' (in Latin if you're fussy). Explain that this gravestone is for Jesus – what could the letters stand for now his grave is empty? (For example, 'really in paradise', 'revisiting 'is Pop'.) Get them to work together and write their suggestions on their cards to take home or display.

To respond to the fact that Jesus is alive

Not there

☑ PREPARATION ☒ BOOK

For each group member put a stone in a box and giftwrap it, first writing on the underside of the lid a clue to where the person's real Easter present (e.g. a chocolate egg) is hidden.

When the group members have found the presents, ask them how they felt on opening their boxes, discovering the clues and then finding the real presents. Point out the parallel with the women's experience on Easter Day – the empty tomb; the message that Jesus was not there but risen; meeting Jesus – the greatest Easter present!

To remember the women's experiences at Easter

Because...

☒ PREPARATION ☒ BOOK

Ask the group members to suggest endings to the sentence 'Because Jesus is alive...' Turn these into a prayer to be used by the whole church by making the statements start with the words: 'Because you're alive...' Begin the prayer with the words 'Risen Lord' and end it with a shouted 'Yes!' Either practise saying the prayer as a group, or write it up on an OHP for the whole church to say together.

To celebrate that Jesus is alive

Ascension

Why celebrate?

Ascension Day is celebrated on the Thursday ten days before Pentecost, and forty days after Easter.

The disciples knew that Jesus would be going away but, as usual, they didn't fully understand. So when Jesus told them to wait for the Holy Spirit and then vanished in a cloud, they were left transfixed, staring into an empty space. He would return, but now it was time for them to move on. Now the disciples would be unable to see Jesus. The evidence that Jesus is alive would be in *their* words and lives.

Ascension Day is an opportunity to celebrate our relationship with the risen Jesus and reaffirm our willingness to be his people on earth – speaking and living for him.

'…why do you stand here looking into the sky?' **Acts 1:11**

Starter pack

Balloon sort (10 minutes)

Beforehand, slightly blow up twelve balloons to the size of a large orange. Write the following statements on the balloons with felt pen:

- Jesus told us about God.
- Jesus made blind people see.
- Jesus taught crowds of people.
- Jesus died on a cross.
- Jesus rose from the dead.
- Jesus was hidden by a cloud.
- We won't be able to see Jesus.
- We will be witnesses for Jesus.
- We will go to the ends of the earth.
- Jesus will come again.
- The Holy Spirit will come to us.
- We will be filled with power.

Give an example of change in your life: how you left things behind and took on new things. Read out Acts 1:6-11. Place two bins, labelled 'past' and 'future' close together. Explain that the words on the balloons refer to the 'past' and 'future' parts of the change the disciples faced. Group members must throw the balloons in the right bins from 2 metres away.

Doing and feeling (10 minutes)

Read out Acts 1:6-10a.

Clear a space. Explain that this is the hillside where Jesus had just disappeared from sight. Ask one group member to stand on the 'hillside' in the role of one of the disciples frozen at that moment. Encourage them to show how that disciple felt. Then ask another to join them. Continue in this way until all who want to take part are in the scene. Read out Acts 1:10b-11.

Sit down and discuss the change that the disciples experienced. They would no longer have Jesus physically with them. Now it was *their* task to show Jesus to the world – telling others and showing his power in their lives. Today, it is still our job to show Jesus to the world as we speak and live for him. Ask the group members to suggest how they can help each other to do that.

Looking up, going on (15 minutes)

Ask each group member to draw faces on three pieces of paper: smiling, frowning and puzzled. These represent the answers 'agree', 'disagree' and 'maybe', respectively. Read the following statements. Each time they must hold up the face which reflects their response. Invite anyone to explain their answer. This could lead into wider discussion.

- Being a Christian would be easier if we could see Jesus.
- I want Jesus to come back soon.
- Sometimes it's easy to talk to people about Jesus.
- We can only live for Jesus if we help each other do it.
- I like hearing what Jesus did when he was on earth.
- I often think about Jesus coming back.

Provide an opportunity for response to Jesus. For example, sing 'We shall stand' (Graham Kendrick).

Idea sack

What's Jesus doing?
☑ PREPARATION ☑ BOOK

Ask the group to speak or write down how they would end the following:
- What I like best about this group is…
- When I see Jesus face to face I'll…
- The church is…
- Oh, you'll never get to heaven in…
- What Jesus is doing now is…

Enjoy their ideas. Read Revelation 1:12-18 – a picture of what Jesus is like now. Explain the picture language, e.g. the stars and lampstands stand for the church, the two-edged sword for Jesus' powerful words. Jesus is glorious – with us safely in his hand.

To be encouraged about Jesus' heavenly glory

Absent friends
☑ PREPARATION ☑ BOOK

Beforehand, write out the names of mission partners linked to the church and group members who have moved away. Give each small group one person or family to write to, telling them about the group, including personal news and asking how they are getting on.

To remember and support those who have moved on

Total celebration
☑ PREPARATION ☒ BOOK

Have a celebration of all Jesus' time on earth from Advent to Ascension with appropriate decorations, music and food for each stage. For example, start by lighting Advent candles. Then put up Christmas decorations, sing carols and eat mince pies. Move on through Jesus' life (Gospel readings and games), death (darkness and a prayer) and resurrection (full-on party with Easter eggs to finish). End with party-poppers or a firework to celebrate Jesus' ascension and promise to return.

To celebrate Jesus

Beautiful but broken
☑ PREPARATION ☒ BOOK

Give each small group a disposable camera and a tape recorder. Send them to walk with a leader around the area. Explain that the aim is to take a closer look at the world where Jesus has left us to serve him. Take pictures or recordings of anything beautiful and anything spoilt.

Play the tapes. Lead into silent or spoken prayer for the people who live in your area. Once the photos are developed display them with the caption 'Beautiful but broken'.

To understand the world where God calls us to serve him

On the hill top
☑ PREPARATION ☒ BOOK

Go to the highest local point, whether hill top or tower block. Read out Acts 1:6-11, asking the group to picture the scene in their minds. Stand silently and focus on the sky for a while, then read verse 11 again.

To think about Jesus' ascension and return

Heaven and earth
☑ PREPARATION ☒ BOOK

Get the church's best musicians to come to the group. Involve group members in planning this worship, to include:
- praise of Jesus reigning in heaven, e.g. 'Jesus, you are the radiance', 'He is exalted'.
- words and songs reflecting how we are left on earth to continue Jesus' work, e.g. 'A touching place' and other songs and liturgy from the Wild Goose Worship Group (see page 64).

To honour Jesus

Pentecost

Why celebrate?

The Jewish festival of Pentecost celebrated the harvest and looked back to the giving of the Law at Sinai. God chose that day to give us something even more amazing to celebrate – his Holy Spirit filling the church with his own presence and power.

The Holy Spirit shapes every part of our lives. Most of all he equips the church for mission. The first effect of his presence was that many people heard about Jesus and believed in him. So our Pentecost celebrations should look outwards – to people of all nations and especially to those who do not know 'the great things that God has done'.

'… each of us… – we hear them declaring the wonders of God in our own tongues!'
Acts 2:8,11

Starter pack

Wait to speak (10 minutes)

Secretly give one group member this message to communicate to the others: *'Say "I'm an aardvark" and you can have some chocolate.'* The messenger must act it out – no speaking or writing. After two minutes, allow the messenger to speak the message. Give chocolate to anyone who responds appropriately. Note that your aim is for the messenger to *fail* at first. Reduce the two minutes if the others are close to guessing.

Read out Acts 2:1-13, first asking the group members to look out for links with your game. Afterwards discuss these. For example: the disciples had something to tell others – Jesus was alive! They had to wait to pass on the message. When they received the Holy Spirit they could speak the message so everyone understood. Many responded to it.

Hearing the Word (10 minutes)

Set the scene: Pentecost was a Jewish harvest celebration and there were lots of visitors in Jerusalem. A few weeks had passed since Jesus rose from the dead, but only his followers knew about it. Now he had gone into heaven. They kept meeting together to pray and wait for the promised Holy Spirit.

Divide into two groups – 'Jesus' followers' and 'the crowd'. Give each group member six pieces of paper with empty speech bubbles on. Read Acts 2:1-13, pausing after verses 2, 3, 6, 8, 11, 13. Each time they should write down an exclamation they might have made at that point in the events (e.g. 'Wow!', 'Help!'). Alternatively, they could speak them.

Explain what happened next: Peter told the crowd all about Jesus – and how he had been raised from death. Read verses 37-41 to see how they responded.

Fruit (15 minutes)

Put the nine qualities of the fruit of the Spirit (Galatians 5:22-23) on labels around the room along with a few red herrings such as 'zaniness', 'style', 'wit', 'charm'.

Explain that everyone who trusts Jesus has the Holy Spirit. The effects of his working in us should show in our lives. Read the following clues. The first person to touch the right label gets some fruit (different each time, e.g. a grape, a chocolate orange, a fruit sweet, a drink of juice).

joy	'Now I know Jesus, I've a deep-down happiness nothing can shift.'
goodness	'I want to do what's right.'
humility	'You first, me last.'
peace	'I forgive you. Don't let's have trouble between us.'
self-control	'So tempting! But, no!'
kindness	'You're struggling. Can I help?'
patience	'I'll keep forgiving him, however many times he does it.'
faithfulness	'I'm hanging in there with you whatever happens.'
love	All the rest together.

Ask: 'If we showed these qualities more, what effect would this have on other people?' Pray for members' relationships with friends, families and each other.

Idea sack

World words
PREPARATION ☑ BOOK ☑

Provide a variety of foreign language dictionaries. Ask the young people in small groups to think up words and phrases of praise to God. Look up the equivalents in the dictionaries and then assemble them into a multilingual chant or prayer. God will understand.

To praise the God of the whole world

Ends of the earth 1
PREPARATION ☑ BOOK ☑

Draw an 8x8 grid (like a chessboard) on paper. Write 'Home' on the four centre squares. Ask the group members to write the following on the grid – one per square:
• places in your area in the squares around 'home'
• other places in your country in the next ring out
• other countries in the outside squares.
When the grid is full, read out Acts 1:8. The grid shows the pattern of how, starting at Pentecost, the disciples would spread the good news of Jesus.

To see how the gospel spread

Ends of the earth 2
PREPARATION ☑ BOOK ☑

This game uses the grid from Ends of the earth 1. Copy the place names on to slips of paper. Each player puts their counter on a home square, then draws three placename slips – one from each category. They must get to those places by rolling a dice in turn and moving the number of squares shown. Counters can turn corners within a move, but not move diagonally. An exact roll is needed to 'hit' the places.

Pray for the spread of the good news, each person naming their places to God.

To see that the Holy Spirit's mission continues

Friendly international
PREPARATION ☑ BOOK ☒

Hold an international event for the group members' families and friends, planned by the young people. Focus on one country with:
• appropriate clothes, food, music, eating implements, dancing
• a talk about the reason for Pentecost
• interviews with guests about different countries
• team quiz – identify where objects come from
• useful phrases in the language learned beforehand.

To communicate the meaning of Pentecost

Closest
PREPARATION ☑ BOOK ☑

Ask the group members to score themselves privately out of five on how close they feel to the others in the group and to God. Look at the following Bible passages. For each one, agree a score for how close God was to his people:
• Exodus 13:20-22; 33:7-11 – In the desert
• Matthew 1:18-23; John 14:8-9 – Jesus
• Acts 2:1-4; Galatians 4:6-7 – The Holy Spirit.
Ask the group members to score themselves again to see if they have changed.

To know God's closeness

Serving instructions
PREPARATION ☑ BOOK ☒

Ask two leaders or members to prepare a sketch in the roles of a practising musician and a boiler-mender in a church.

At first both work quietly. The noise builds up as they try to drown each other out. They stop and argue about who is most important to the church. The boiler-mender KOs the musician. A group member reads 1 Corinthians 12:4-7. The boiler-mender shrugs and drags out the musician.

To show that the Spirit's gifts are for service

Harvest

Genesis 8:20-22; Matthew 6:25-34

Why celebrate?

In agricultural communities, harvest celebrations can still mean a great deal. In urban areas, they often seem second best. But the regular appearance of food in fields, in supermarkets and on our tables is a fulfilment of God's promise to us all. Harvest keeps coming round because of God's patience and mercy (Genesis 8:20-22). It's an undeserved gift.

Jesus challenges us to live and give without counting the cost or worrying about the future. We need to put God's purposes above our own physical wants and needs. Harvest is an opportunity for our young people to understand that all they have comes from God, and to praise and thank him genuinely.

'But seek first his kingdom and his righteousness, and all these things will be given to you as well.' **Matthew 6:33**

Starter pack

What's wanted? (10 minutes)

Make sets of cards showing these words: TV, CD player, bike, new clothes, magazines, coolest trainers, books, big meals, spectacles, water.

Give each group member two cards. Explain that the aim is to get items they want by swapping. If one person says 'swap' to another then they *must* exchange a card, each choosing which of their own to hand over. Stop after two minutes. Ask each person to select one of their cards and try to justify why that item is essential to life.

Give each group of three or four a complete set of cards. Ask them, in silence, to arrange the cards in a league table of importance, taking turns at adding one card at a time. When all the cards are placed, each person may change the position of one item. At the end allow (perhaps heated) discussion.

Don't worry – be happy! (10 minutes)

Before the session, collect items linked to words in Matthew 6:25-34, e.g. clothes store card ('clothes'), hayfever tablets ('wild flowers'), Bible ('what God requires'), *Bird's* custard. Bring the items along to the session.

Place the objects in the centre. Read out the passage. Each time someone spots a link with an object they must grab it. Offer a prize for the corniest link.

Ask the members, in small groups, to look back through these verses and write down a few key phrases to sum up what Jesus said. Get them to design and make new wrappers for the food items from the game, putting across these key ideas. Stack them into an effective display somewhere visible.

Blow the facts! (15 minutes)

Play in teams. Each starts with eight balloons. Ask the following questions. When all the teams have guessed, announce the correct answer. For each wrong answer, burst a balloon. The team with most balloons at the end wins. (Correct answers are starred *.)

1. How often do children in Zambia have breakfast cereals?
   a) four times a week
   b) once a week
   c) never*

2. How do numbers of births and deaths compare worldwide?
   a) equal
   b) more births*
   c) more deaths

3. Roughly how many people in India can read?
   a) one in ten
   b) one in three*
   c) nearly all

4. What is the average weekly wage in Brazil?
   a) £21*
   b) £57
   c) £126

5. How often does a typical Nepali family eat meatless lentil stew?
   a) once a week
   b) once a day
   c) twice a day*

6. How many doctors are there in Haiti? One for every...
   a) 80 people
   b) 900 people
   c) 11,000 people*

7. How many tubes of sweets could a typical man in Liechtenstein buy with his day's wages?
   a) 500*
   b) 300
   c) 68

8. How many people in Costa Rica live in poor crumbling cities?
   a) one third
   b) one half*
   c) all

Ask the group to look through information from Christian development agencies and plan a practical way of making a difference to such unequal situations, e.g. by giving money, praying, writing to national leaders or raising interest more widely.

Idea sack

In tents

☑ PREPARATION ☒ BOOK

Hold a sleepover in tents linking to the Jewish harvest festival – the Feast of Tabernacles or Shelters. Ensure good supervision and separation of lads and girls. As they feast, read Leviticus 23:33-44 to give the background. Have a bit of structured activity – but not too much, e.g. videos or other activities on these pages. For more background read Deuteronomy 8:10-18. When our lives are comfortable we can forget God, too. Chat about how we can combat that danger.

To remember our dependence on God

Exotic tastes

☑ PREPARATION ☑ BOOK

Show the group a number of exotic fruits and vegetables. Each member or pair should write down what they think it is called (or invent a name for it) and guess how it should be eaten. Finish by sharing them all. You could also provide samples at the back of church after a harvest service.

To enjoy the variety of God's provision

Food auction

☑ PREPARATION ☒ BOOK

Prepare small plates of different foods, nice and nasty: e.g. apple, cheese, cake, cup of water, mushy peas, cold rice. Have around one and a half times as many items as group members. Put different amounts of toy money from £50 to £500 in envelopes.

Give out the envelopes at random then auction the plates of food. Afterwards read Genesis 8:20-22. Our supply of food is a sign of God's love – not something we deserve. Ask how that might affect our attitude to those who have less.

To explore world inequalities

Word harvest

☑ PREPARATION ☑ BOOK

Hold a session when all the young people are encouraged to bring along favourite items of writing. They may choose passages from the Bible, books or poems, or even teenage magazines. Allow them each to read a section (or nominate a friend to do so), and ask them briefly to explain why the writings mean something special to them. Make sure the leaders do the same, too. You could let them bring tapes or CDs instead.

To build group relationships

Looking up

☒ PREPARATION ☒ BOOK

On a clear night, wrap up warm and all lie on your backs and look up at the stars. Provide groundsheets to lie on. Encourage the group members to comment on what they can see. Then read out Psalm 136:1-9 with the young people saying the response 'His love endures for ever.' Finish with a time of quiet.

To be amazed at the creator God

Ad gap

☑ PREPARATION ☒ BOOK

Spread out some interesting magazine adverts. Invite group members to pick up and comment on those they like or dislike. Spread out amongst the adverts some pictures of people living in poverty from aid agency magazines. Again, invite comment. Fill in some information about the individuals pictured. Ask the group how they react to the combination of items.

Pray briefly, committing the ideas and views expressed, and the people pictured, to God – but without repeating them all.

To pray about world inequalities

All Saints

KEY BIBLE PASSAGE   1 Peter 2:4-12

Why celebrate?

We are. All saints, that is. As Peter puts it, we are a holy nation – a people who belong to God, set apart for him. That is certainly something to celebrate.

All Saints celebrations on 1 November have been restarted in many churches to provide young people with an alternative to Hallowe'en events. God has called us out of darkness into his own light. We have something enormously positive to celebrate instead – God has made us all his.

All Saints Day is a good time to encourage young people with a glimpse of the big picture. We are one with God's people throughout history. We can learn from Christians in the past: inspiringly holy and also reassuringly imperfect.

'But you are… a holy nation, a people belonging to God, that you may declare the praises of him who called you out of darkness into his wonderful light.' **1 Peter 2:9**

Starter pack

Dark to light (10 minutes)

Write out the following individual words on cards: lost, found; enemies, friends; empty, full; lies, truth; captive, free; fear, hope; poor, rich; blinded, seeing; evil, holy; night, day; distant, near; dead, alive.

Secretly, give each pair of group members up to six cards, splitting up the matching pairs of words (e.g. one pair has the word 'lost'; another pair has the word 'found').

Ask each pair privately to draw something to represent each word (like in *Pictionary*) – on separate pieces of paper. Then, as a group, spread all the pictures on the floor. Explain that they represent opposite pairs. All work together, with no talking, to pair them up.

Check the pairings are correct. Explain that they are all words used in

the Bible to show the difference between being a Christian and not. Ask the group members for any examples of such contrasts they have seen in themselves or others.

Read angles (10 minutes)

Try the following ways of presenting 1 Peter 2:9-10 together. Alternatively, get a small group to work on each one and present it to the rest.

1. Write a long list of Christians: people you know and famous Christians of today or the past. One person should read out 1 Peter 2:9-10, pausing after each phrase while the others say a selection of the names.
2. Darken the room. Use lights, torches, OHPs, etc., to make a spectacular (and safe) light display. Read out 1 Peter 2:9-10. Where it speaks about life with God, switch on the lights. Where it's about life without God, switch them off.
3. Work out simple dramatic movements to go with 1 Peter 2:9-10. At the same time others should reword the verses so they are addressed to God (e.g. 'We are your chosen race...') Put the two together.

At the end invite comment. You could explain that these verses do not assume a dramatic conversion is always necessary. They're about what God has done for everyone who believes in Jesus.

Three lives (10 minutes)

Get three church members to appear as:

- a famous Christian from the past (not in the Bible)
- someone from your own church's past
- him or herself.

Let them choose people they know about or can easily research. Use a TV chat show format to interview them with main questions agreed in advance. You might cover:

- How did you become a Christian?
- What role(s) have you had in the church?
- How was the church different in your time? (First two only)
- What would you most like to say to young Christians today?

Pray, thanking God for the lives of those three people.

Idea sack

Saintly feast

☑ PREPARATION ☒ BOOK

Put on a formal meal for one section of the church, e.g. older people, the church council or the children's group leaders. Give it an 'All Saints' feel and plan how to make the guests feel really special. Send formal invitations. Have place labels, e.g. 'Saint Ethel'. Choose food which the members can help prepare. Others should dress formally to serve. Other 'saints' in the church might contribute financially.

To serve and honour others in the church

Hot potatoes party

☑ PREPARATION ☒ BOOK

Invite group members to suggest 'hot potato' questions they would like answering about the church, living as a Christian, the Bible or the world. Hold an outdoor party, perhaps with a bonfire, serving baked potatoes and a choice of fillings. Invite church leaders along. After the food, let the group members ask the panel their questions. Plan in advance which to use and group them in themes. Finish with songs and, perhaps, fireworks.

To ask real questions about faith

Footprints of the saints

☑ PREPARATION ☒ BOOK

Get the group members to take sets of footprints from as many church members as possible. Place long strips of wallpaper or coloured frieze paper on newspaper and provide trays of paint (experiment for thickness). Get a pair of footprints from each person. Get them to sign their name under the prints. Display them with the caption 'Footprints of the saints'.

To affirm church members and build relationships

Slipping halos

☑ PREPARATION ☒ BOOK

Give each group member a halo to balance on their head, e.g. a frisbee, pan lid, or hub cap. Time how long they can keep a balloon in the air without any halos being held or falling off. The same player must not hit the balloon twice in succession.

Afterwards ask: 'How easy is it to be a saint?' Look together at Colossians 3:12-17 and ask: Who is it that makes people saints? How can we help each other live up to that title?

To help each other live as saints

Pics and facts

☑ PREPARATION ☒ BOOK

On a board, Blu-Tack photos of church members (not just those you get on best with) and just a few pictures of famous Christians from history. Cover each with a numbered card. Ask group members to call out numbers. Remove the card. The whole group (or a team) have to name the person and give one piece of information about them.

Thank God and pray for the people pictured, saying their names aloud.

To get to know the wider church present and past

Halo hurl

☒ PREPARATION ☒ BOOK

Two teams each choose a 'saint'. The saints stand on chairs at opposite ends of the playing area. The teams must get 'halos' (rings of thick card 20cm across) to their saint by throwing them from person to person. They must not run while holding a halo. They can only steal halos by intercepting passes. Start with each team holding three halos. The holiest team is the one whose saint finishes wearing the most.

To build relationships

Remembrance

KEY BIBLE PASSAGE — *Psalm 40:1-5*

Why celebrate?

Remembrance Day is marked on the second Sunday in November, recalling the signing of the armistice at the end of the First World War at 11 a.m. on 11 November 1918. Since then we have had many other conflicts to remember and reflect on.

The writer of Psalm 40 is exhilarated by God's rescue from a dire situation. He praises God, speaks of what he has done and affirms his desire to do God's will. In times of world or personal crisis, we too should put our trust in God. War time has given us amazing examples of such trust, even from the horrors of the concentration camps. And when the crisis is over we need to praise him for what he has done. In looking back at past wars, we need to give thanks for the peace in which we live, and also affirm our decision to use that peace for God's purposes – to live and speak for him.

'I waited patiently for the Lord... He lifted me out of the slimy pit...' **Psalm 40:1-2**

Starter pack

Milestones (10 minutes)

For each group of up to seven members, a leader should bring about ten items which mark different stages of his or her life, e.g. photos, exam certificate, birthday card, school book, something they inherited, a baby's first clothes. Ask the group members to look at these and to try arranging them in chronological order. (Don't make it too easy.)

Afterwards explain the items. Ask each group member to suggest two items they would have brought to represent their own major life events.

Explain that it is good to look back at the past – to thank God for what he has given us and done for us; to remember the crises from which he has rescued us. On Remembrance Day we also look back further to remember and thank God for things that are beyond our experience. Ask the group to imagine holding one of the objects they have named and to thank God silently for what he has done in their lives.

Out of the pit (15 minutes)

Talk in pairs or threes of the members' own choosing about the worst danger or fear they have experienced. Point out that in peace time it's easy to forget the terrible realities of war. Ask the group members to talk about any recent examples of conflict around the world which they have heard about.

Show a brief video clip from a wartime newsreel or a feature film about war, preferably something which shows the fear and severity without being completely shocking. Freeze the video and read Psalm 40:1-8.

If possible, give an example of someone who has shown great trust in God in wartime. For example, read or show a clip about Corrie Ten Boom's experiences in Ravensbrück concentration camp from the book or video of *The Hiding Place*.

Remembering (10 minutes)

Get hold of the Roll of Honour for your church or locality, naming those who gave their lives in the last two world wars. Alternatively, visit your local war memorial.

Pray: 'Thank you, Lord, for peace.' Read out the names on the Roll of Honour one by one. Repeat the prayer at the end.

Idea sack

Back chat
☑ PREPARATION ☒ BOOK

Ask the group members and leaders to find out what family members did in the two world wars and to report their stories back to the group, bringing along any relevant photos or objects. At the same time ask them to find out any stories about the Christian faith and church activities of the previous generations. Again they should bring any items that will help them report back.

To make connection with real wars and Christians in the recent past

Peace memorial
☑ PREPARATION ☒ BOOK

Get the group members to make something like a war memorial but about their life in the group. Include: a drawing of the group in action; the words 'In memory of those who gave their lives to (*name of group*); the year; their initials and surnames; a Bible verse of their choice. The verse and the drawing should show what they would like to be remembered for. For the finished article use a board painted grey or a large sheet of stone-coloured sugar paper.

To consider how to use peacetime

Peace zone
☒ PREPARATION ☒ BOOK

Take the group for a walk to somewhere peaceful and quiet. Remind them of the reason for Remembrance and then encourage them to sit, walk or stand and think through what they can do in small ways to encourage peace, beginning with their attitudes and relationships.

To think through practical ways to peace

Time trip
☑ PREPARATION ☒ BOOK

Take a quick trip to a local site connected with a battle or conflict from any period of history, e.g. a battlefield, castle, site of bomb damage or scene of riots. While there, explain what happened. Pray silently for peace in the world. Go on to a place with a Christian past, e.g. a hospital or scene of open-air preaching. Explain the significance of the place and pray silently for the church as a light for the world.

To pray for peace and for the church

Two sides
☑ PREPARATION ☑ BOOK

Gather news information on a current scene of conflict on video, tape or magazine and newspaper cuttings. Divide the group into two. Ask each half to represent one side in the conflict. Let the groups use the material to find out what's going on and then briefly report what's happening from their side's point of view.

Share food and drink together. Read Isaiah 25:6-9 then have open prayer (one-word or longer prayers) for the conflict situation you have explored.

To pray for God to bring peace

Two minutes
☒ PREPARATION ☒ BOOK

Hold a two-minute silence. Introduce this by saying that it is a time for us to remember about people who died in the world wars and other conflicts and to pray. Afterwards, ask the group members to say honestly what they thought about – even if it's totally irrelevant to war. Write these down. Rearrange the ideas into a poem: 'In the two minutes' silence, I thought about… (*list of thoughts*) … in the two minutes' silence.' Publish it in the church magazine.

To be honest about the difficulty of Remembrance

New Year

KEY BIBLE PASSAGE   *Matthew 11:28–30*

Why celebrate?

For some, New Year is a time for great parties. For others, it feels flat. Either way, it is a natural point for looking back to the past year and forward to the next. If we ignore God, that easily turns us in on ourselves: to self-satisfaction or despondency about the past, empty optimism or stress about the future. We need to see that the past and future years are in God's control and to celebrate what he has done, is doing, and will do.

Jesus called people to bring their burdens to him and swap them for another load that he gives. He calls our young people to trust him with the issues from the past year and to take on his priorities for the next.

'For my yoke is easy and my burden is light.' **Matthew 11:30**

Starter pack

Look back (10 minutes)

Place the names of all the months around the room. Make a list of twenty events from the last year – in the group and the wider world. Call them out one at a time and invite the young people to go to the month in which they think it happened.

Sit in a circle. Ask each person in turn to share good times and hard times from the past year. Pass round a clock to show whose turn it is to speak. If possible, start it off with your own reminiscences. Allow them to pass the clock on without speaking.

Where's God when...? (15 minutes)

Write out the following Bible verses on different coloured paper. Cut them up and put each individual word into a separate balloon. Blow up the balloons.

- I am with you always. (Matthew 28:20)
- So do not worry... (Matthew 6:31)
- ...you, O Lord, are exalted for ever. (Psalm 92:8)
- ...he turned to me and heard my cry. (Psalm 40:1)
- Come to me, all you who are weary... and I will give you rest. (Matthew 11:28)

When you say 'Go!' the group members must burst the balloons and arrange the words of each colour to get the verses without using Bibles.

Discuss the verses all together or with a small group gathered round each one. Ask:

- What do the verses mean?
- How easy is it to trust them in good and bad times?
- What do we need to do about them? (e.g. How can you call on God?)

Use your background preparation on the verses to help.

Looking at the future (10 minutes)

Write out the months 'January' to 'December' along a roll of paper, e.g. the back of wallpaper. Invite the group members to invent their 'Fantasy Year', by suggesting what they would like to happen in each month, e.g. 'City win the League', 'Extra ten weeks of holiday', 'Get grade A in all exams'. Either write their ideas up under each month or get them to do this.

Next, challenge them to suggest what God might like to see happen in the next year – in their lives, in your group, in the world. Write these up in a different colour, not in any particular month.

Explain that New Year resolutions are only worthwhile if our hopes are based on what God wants for us. Ask the group to sit or walk quietly for three minutes, thinking about changes God might want them to make in their lifestyle or attitudes.

Finish by praying for God's help for all of you in the coming year.

Idea sack

Group calendar

☑ PREPARATION ☒ BOOK

Introduce a group calendar or year planner for the year ahead. Write in any future events and dates of terms as you plan them – marking them with coloured stickers. Allow group members to add in their own 'life events', e.g. moving home, appearing in a school play. At the end of each month, get a group member to write in what you have learned about God. Look back at these from time to time as a reminder.

To see what God has done through the year

Clean out

☑ PREPARATION ☒ BOOK

Have a fun working party to clean the church building. Provide great food and have the group members' choice of music playing. Encourage as many of the young people as possible to join in. Let them go home and change before meeting at someone's home for a video in the evening.

To build relationships and serve the church

Note to self

☒ PREPARATION ☑ BOOK

Invite the group members to write postcards to themselves – to be opened in six months' time. They should include:
- how I'm feeling
- what I want to have done or changed in six months
- something I'd like to remember about God.

Seal the cards in named envelopes, put them away and make a note in your diary. When they open them discuss whether reading their message is encouraging or discouraging. Have they seen God at work? Pray for them – looking back and forwards.

To reflect on life and see God at work

On the fringes

☒ PREPARATION ☒ BOOK

Discuss and make a list of all those who live in the area who may feel rejected and on the fringes of the community, especially at New Year. These may include children, lonely people, older people. Discuss what the group members could do to help one of these groups – visiting older people, babysitting, fund-raising, etc.

To help those in need

Making history

☒ PREPARATION ☒ BOOK

Take a few photographs of the group in poses of their choice. Make sure their faces are showing. When they are developed, select one and put it in an album or frame with space for more pictures. Start a tradition – add a new photograph each year as a record of the past and an aid to praying about the people shown. Decide to record the rest of your year's activities with occasional photos, too.

To have a record of what happens in the group

Perfect prediction

☑ PREPARATION ☒ BOOK

Cut items from weather forecasts, bus timetables, horoscopes, articles about the year ahead, pools forecasts, diaries, TV guides or other material about the future. Give each small group twelve predictions to arrange in order of how much they would trust them.

Ask what makes predictions reliable. Read Joshua 1:5. Point out that however unpredictable the year may be, God's part is certain – we can be completely confident he will be with us through it.

To face the year ahead with confidence in God

Births and birthdays

KEY BIBLE PASSAGE (*Psalm 139:13-16*)

Why celebrate?

The birth of a baby usually leads to great celebration. The date is then treasured – each person's birthday is the day when those around them acknowledge how special they are.

The writer of Psalm 139 speaks of God's remarkable creation of him as an individual. As the maker, God knows absolutely everything about him. The response from the writer is wonder and praise. We need to include these aspects in our birth and birthday celebrations.

With group members you might want to celebrate the birth of a baby in one of their families, or in the church. You could also have an annual 'official birthday' when you celebrate the lives of all the group members. And marking individual birthdays through the year is a brilliant way of showing how we and God value each person for who they are.

'You created my inmost being... I praise you because I am fearfully and wonderfully made.' **Psalm 139:13–14**

Starter pack

Unique? (10 minutes)

Ask the group to get into clusters according to the following categories. Each time give those in a cluster of one (standing alone) a chocolate.

- born in (*local area*)/ elsewhere in (*your country*)/another country/don't know
- month of birthday
- eyes mainly green/brown/blue/grey
- day of week of birthday this year (or 'don't know')
- wearing any red today/not
- share identical DNA (only identical twins will be together)
- age in years
- number of letters in full name.

Ask the group members: Would you rather blend in or stand out – be

like everyone else or be different? Point out that God is creating life every second of every minute of every day! He makes us all unique (even identical twins). Sometimes it can be tough being different from other people. But each person he creates is infinitely precious to him.

Humanoids (15 minutes)

Work in groups of two to four. Give each group different materials and challenge them to make a human being. Groups might have newspaper; old boxes; sticks; bread; bottle tops; carrots. Provide sticky tape, glue and scissors for all the groups. Each group should give their person a name. Praise their efforts and perhaps award a prize for the most human-looking.

Place the creations in the middle. Read out Psalm 139:13-14. Ask group members to suggest what they think is most 'strange' or 'wonderful' about human beings and the way God has made us. How are we different from their own humanoid creations?

Praise God for each other. Start and end with the words: 'We praise you, Lord, because we are wonderfully made.' Say the name of each group member in turn. Each person could say the name of the person on their left.

Now I can! (10 minutes)

Make an enormous paper cut-out of a cake. Divide it into three areas and label them '0–5', '6–10', '11–15'. Invite the group members to write or draw in each space things which they learned to do at that age. The aim is to cover the space as quickly as possible (so write big). Pause and spend a couple of minutes looking at the sheet. Play a worship song which speaks of God's love for all of us.

Give each group member and leader a small cake with a lighted candle on. Explain that today is an official birthday for all of you. Remind the young people that from birthday to birthday they change, and God is with them all the time helping them as they grow and develop.

Idea sack

Birthday chart

☑ PREPARATION ☒ BOOK

Keep a chart of group members' birthdays on the wall of your meeting place to help you all celebrate those dates. It could also be a reminder to follow up any members who have not been seen for a few weeks. Encourage the group members to share in this – contacting those people tactfully to find out why they are not coming and feeding back information to the group for prayer or action.

To celebrate birthdays and take responsibility for each other

You're a star because...

☑ PREPARATION ☑ BOOK

Prepare star badges at least 10cm big from card of different bright colours for all the group members and leaders. Ask each person to write their name in the centre of the badge. Stick them up around the wall and invite everyone to write on four badges something they like about that person which makes them a 'star'. No badge should have more than four comments on – so everyone has the same number.

To build relationships and help group members see their own worth

Cool cards

☑ PREPARATION ☒ BOOK

Ask the group members to cut out bits of magazine pictures and stick them together on plain postcards to make funny images on the theme 'birthdays and celebrations'. Agree suitable captions together. Either sell the cards for group funds or keep a stock to send from the group to other people, e.g. those who have moved away, mission partners, church leaders, all the over-60s in the church. They could keep a birthday book to help with this.

To celebrate birthdays of others

Trad birthday

☑ PREPARATION ☒ BOOK

Make a tradition of marking each group member's birthday at the meeting on or before the day itself. You could get them to agree how this will be done, e.g. have some kind of cake or bun with a candle in it (different each time), say a birthday chant (invented by the group) and pray for them.

To affirm every group member

Nappy birthday

☑ PREPARATION ☒ BOOK

To celebrate the birth of a baby in the family of a group member or leader or in the wider church, send a disposable nappy instead of a card. Write a message on it and get all the members to sign it. As a gift, get a babygro (new or second-hand) and ask the group members to do a design on it with fabric paints.

To celebrate a birth

0 to 90

☑ PREPARATION ☒ BOOK

Randomly share out cards showing ages from 0 to 90. Have a party where each person must dress the age on their card. Party ideas:
- a mix of food, music, videos, games for babies to 90s
- picture wall quiz – guess celebrities' ages
- sort Bible references in order of age of the people used/blessed by God – Luke 1:44, 1 Peter 2:2, 2 Kings 22:1, Mark 5:42, Genesis 37:2 (Joseph), Luke 3:23, Acts 7:23, Genesis 25:26 (Isaac), Genesis 17:24, 1 John 5:11
- Pray and read Isaiah 46:3-4.

To celebrate God's love for us at all ages

Start of the school year

Proverbs 3:5-6 and other verses

Why celebrate?

Facing a new school year, group members may experience excitement or fear, uncertainty or confidence. But all those feelings can be misleading. Proverbs is packed with wise advice for life: helping us get a right view on ourselves, others, God and the world around. It all starts with trusting God in every situation – looking to him to direct us, rather than relying on our own view of the situation. For our young people that applies in every aspect of school life: classes, homework, friendships, making choices, relating to teachers.

By marking the start of the new school year we can acknowledge God's concern with whatever they are going through and look to him for direction and protection. We can help them support each other as well as giving our own care and encouragement.

'In all your ways acknowledge him, and he will make your paths straight.' Proverbs 3:6

Starter pack

A to Z of school (10 minutes)

Beforehand, write a list of 'things at school' starting with the letters A to Z, e.g. 'art', 'books', 'classrooms'… Go for obvious ideas. Make a second list of 'words to describe God', e.g. 'amazing', 'brilliant', 'caring'… ('eXceptional'?).

Announce the topic 'things at school'. Go round the group working through the alphabet. Give a point (or blow a hooter) each time someone gets a word on your list. Allow them to help each other. Alternatively, allow anyone in the group to call out for each letter. Repeat the exercise with: 'words to describe God'.

At the end, make the point that the God they have described cares about all the school things in the first list. School matters to him because the people matter to him. What we do there matters to him, too. Read out Proverbs 3:6.

Hot tips (15 minutes)

Write out the following six verses in full on separate cards: Proverbs 12:25; 15:19; 17:17; 21:23; 23:12; 29:25. On a seventh card write: 'Don't get mad – get even!'

Explain that you have six hot tips from the Bible for the start of the school year – and one fake. Read out the seven cards, asking the group members to spot the one which is not from the Bible (the fake card actually says the opposite of Proverbs 20:22). Rip up the fake card when they have guessed it.

Divide into groups of two to four and give each group a card. Ask them to think of an imaginary incident at school which would show the truth of their saying. They could act these out if they want.

You could use the feedback to start wider discussion of particular concerns, e.g. building good friendships or coping with what others think of you.

School stuff (10 minutes)

Place school-related items on the floor, e.g. a book, a pad of paper, a tie, a bus ticket, a computer mouse, a flipchart pen, a test tube, a sandwich. (You could ask group members to bring these.) Ask each young person to think of something they are looking forward to in the coming school year, and one thing they are fearful of. They could talk about these if they want.

Lead a prayer time. First commit their hopes and fears to God. Then use the objects to spark off prayer ideas. As you pick up each object in turn, name a related area for prayer (e.g. bus pass – 'Let's pray about travel to school'). Leave a brief silence for each. With a confident group, members could pick objects and name topics.

Idea sack

School board
☑ PREPARATION ☑ BOOK

Put up a pinboard with Proverbs 3:6 along the bottom. Get the young people to write or draw ideas for prayer about school. Pin them up. They might be general ('homework') or specific ('Jim in our class is being bullied'). Include items to thank God for, needs and events. Use the board frequently for prayer. Look to see where God has been at work in those situations.

To pray consistently for school situations

Tip
☑ PREPARATION ☒ BOOK

Play a version of the Waddingtons game *Pit*. Make your own cards with the following suits: wisdom (100 points), knowledge (90), friends (80), good marks (70), homework (60), bus tickets (50). Have a 'cool' card instead of the 'bull' and a 'bully' to replace the 'bear'. You will need a set of cards for each group of four to six. (Toy/games shops sell packs of blank playing cards.) The suits show things collected at school. The best is wisdom – God-sense. Play to the usual rules.

To reinforce that wisdom is the best aim

Landmarks
☒ PREPARATION ☒ BOOK

Ask each group member to think of three points on their route to school. Suggest that as they pass them they do the following in turn:
- remember that Jesus is with you
- pray by name for others you will see
- tell God about yourself and how you feel.
The initial letters JOY will help them remember this. Go over it a few times.

To pray themselves to school

First day reactions
☒ PREPARATION ☒ BOOK

Work in groups of three to five. Assign these roles: older student, head, new student's parent, (new student), (teacher). Announce the following situations that happen before school on the first day. Each person must voice their character's reaction.
- The head backs into the parent's car.
- The teacher falls flat in some mud.
- The older student jumps out on the head by mistake.
- The new student saves the school by putting out a fire.
Discuss whether schools are places of conflict or teamwork. What part should Christians play?

To think about school relationships

Last rites
☑ PREPARATION ☒ BOOK

On the last afternoon or early evening of the holidays, invite the group members to celebrate the 'last rites of summer'. Do what they enjoy, e.g. games and food in the garden. Get them to agree to do all their preparations for school beforehand. Guarantee to parents that you will get them back by 8.30 p.m. Encourage them to have an early night.

To make a good transition from holiday to school

Where I am
☑ PREPARATION ☑ BOOK

Give each member a copy of the following verses written out: Psalms 9:9-10; 13:1-2; 23:1; 25:4-5; 31:9; 51:1-2; 64:1-2; 100:1-2; 130:5; 136:1. Invite them to choose which verse reflects most closely how they are feeling about the start of the school year. Some may like to say which they chose and why. Encourage them to keep the sheets and to look up the rest of their psalms.

To get help from the Psalms in praying

Tests and exams

KEY BIBLE PASSAGE   *Psalm 40:12-17*

Why celebrate?

Our young people are under more pressure than ever to achieve. Faced with tests and exams, they will feel stress, self-doubt and a mixture of fear and nervous anticipation. Some may feel isolated, others totally overwhelmed.

The writer of Psalm 40 was engulfed by troubles, by his own failure and by his enemies. But he called out to God. He knew that even in the middle of his suffering God would help and rescue him. Our young people need to know that they are not on their own. They can always call on God. We can support them through their struggles, while at the same time providing a balanced perspective – showing that their real value under God is unchanged, whatever results they manage to achieve.

'Be pleased, O Lord, to save me; O Lord, come quickly to help me.' **Psalm 40:13**

Starter pack

Blindfold Jenga (10 minutes)

Play Jenga or a similar game, which involves removing individual building blocks from a stack of blocks without making it collapse. Start with one normal round of the game. Then divide into pairs. Blindfold one of each pair. These blindfolded people must play in the usual way – only touching the stack of blocks with one hand. Their partners may help by guiding their hands and giving instructions, but must not touch the blocks. The players do not have to place the removed blocks on top of the stack. When the stack falls, swap over the blindfolds for the next round.

Afterwards ask the group members how it felt being a player or helper. Then ask: 'When you're struggling with tests or exams, who can help or support?' Discover what help the group can give. Make the point that most of all we need God's help. He will always be with us. Read Psalm 40:12-17 as an example of prayer in a tough situation.

Under test (15 minutes)

Beforehand, put the names of the following Bible characters round the room. Write their successes on thumbs-up-shaped cards (two for each character). Adapt these for the Bible knowledge of the group.

Joseph	his dreams came true/became a top guy in Egypt
Moses	brought up by a princess/led the Israelites to the Promised Land
David	defeated a big head/became ruler of Israel
Esther	won a beauty contest/saved the Jewish people in the Persian Empire
Nehemiah	had a top job for the king of Babylon/rebuilt Jerusalem
Mary Magdalene	healed by Jesus/first to find Jesus had risen from the dead
Stephen	one of the apostle's helpers/had a vision of Jesus
Peter	called 'Rock' by Jesus/preached at Pentecost – 3,000 became Christians

Challenge the group members to stick the cards on the appropriate name labels.

Point out that many people in the Bible seem great successes, but they went through great struggles. Divide up and give each small group one character and the following Bible reference: Joseph (Genesis 37:23-28; 39:1-3), Moses (Exodus 17:1-7), David (1 Samuel 23:14-18), Nehemiah (Nehemiah 4:7-13), Esther (Esther 4:15 – 5:3), Mary Magdalene (Mark 15:33-41), Stephen (Acts 7:54-60), Peter (Acts 12:1-10). Ask them to discover:

- What difficulties did this person face?
- What did they do and what did God do?

Listen to the groups' findings. Conclude that all these people discovered that God was with them in their difficulties – though the problems did not *always* get sorted out quickly or easily.

Calendar (10 minutes)

Put together a calendar of dates when group members have tests or exams. Agree to pray for each other on those dates. Ask the group members to suggest sentences or phrases that they could use to pray for themselves or each other in exams. Include Psalm 40:13. Make copies of the exam prayer diary including these ideas.

Idea sack

Me too!

☑ PREPARATION ☒ BOOK

Ask the group members to think how they are feeling – whether facing exams or not. Provide lots of magazines and ask them to cut out either:
- a person who looks like they feel, or
- a scene which looks like what they are going through.

Ask some or all to talk about their pictures and why they chose them. Finish by praying for all the group.

To reflect on and share situations and feelings

Exam questions

☑ PREPARATION ☒ BOOK

Have a panel of three: a group member, a mature older teenager and someone who's just left university. Ask the panel ten general quiz questions as a warm-up. Then have serious questions about tests and exams. Get the group to think these up beforehand, e.g.:
- What are your best and worst exam experiences?
- Does being a Christian make a difference to you in exams?
- What's your top tip for people taking exams soon?

To think through a Christian approach to exams

First post

☑ PREPARATION ☒ BOOK

Send an encouraging card to each group member taking tests or exams – to reach them on the day before their first one. If you've done 'Me too!', use the pictures chosen for the front of each person's card. As a group, use 'Writing Icing' to put encouraging messages on chocolate biscuits to give to those doing exams.

To encourage group members taking tests and exams

Screen test

☑ PREPARATION ☒ BOOK

All watch a short video clip from a TV drama. Ask the group members to watch it again and devise ten questions to ask the leaders, e.g. 'What colour was the cat?' After each question, all the leaders should write down their answers and then announce them in turn.

Talk about the difficulty of remembering things under pressure. Swap ideas for exam revision. Point out that the most important thing to remember is that God is there with you.

To help each other with revision techniques

Nightmares and dreams

☒ PREPARATION ☒ BOOK

Give this scenario: 'You dream that you've just woken up on the morning of an exam.' Ask the group members to continue the dream, passing the storyline from one to the next when they want to. Make the first storyline a nightmare, the second a wonderful dream.

Ask the group members to think silently of the worst and then the best things that could really happen in their exams. Read Philippians 4:6-7 and entrust those fears and hopes to God.

To face up to hopes and fears and trust God for the future

Safe

☑ PREPARATION ☑ BOOK

Write out the second part of Romans 8:39 in the centre of a large piece of paper, e.g. several strips of wallpaper taped together. Spread this on the floor and invite the group members to draw visual images to show what the verse says and what it means to them. Provide a good range of marker pens or paint and decorator's brushes. Talk through their responses together.

To recognize security in God's love

Change of school

KEY BIBLE PASSAGE   Isaiah 43:1–5

Why celebrate?

It is a real challenge to start again back at the bottom of the pile. The familiar buildings, friends and teachers have gone and everything is unknown. There are new timetables to understand and routines to discover. Changing schools can be very scary as well as having the excitement of a step of growing up.

Isaiah told the Israelites that the God who made them and loves them would protect them – even through the apparent disaster of exile in Babylon. In comparison, starting at a new school seems a small trial to go through! But God's love is no smaller. He won't leave them to struggle through alone. His promise, again and again in the Bible, is to be with his people.

Newcomers, especially at the youngest end of school, are often looked down on. By contrast, our groups can demonstrate God's love by supporting and sharing experiences with those starting at new schools.

'When you pass through the waters, I will be with you.' **Isaiah 43:2**

Starter pack

Challenge check (10 minutes)

Write the following topics on cards, with the numbers in brackets on the backs. Make enough cards for each member to have one – with duplicates if necessary. Place a card on the floor in front of each member, with the number facing up. Take it in turns to roll two dice. When a number between 6 and 12 is rolled, the person or people with that card must either say something about it or act it out. If a number is thrown a second time ignore it.

- Making a new friend (6)
- Getting lost in your new school (7)
- Arriving late at a lesson (8)

- Trying out some new school equipment (9)
- Getting too much homework (10)
- Getting home after your first day at a new school (11)
- Getting back your first piece of work from a teacher (12)

Chat at the end using these ideas to draw out more about the group members' experiences of changing schools.

Deep water (15 minutes)

Split into two teams with everyone in sight of a Bible (the same version) open at Isaiah 43. Explain that you will read words picked from verses 1-5. The first team to call out the correct verse number gets a point and must read the full sentence or line.

- the Lord who created you (verse 1)
- your Saviour (verse 3)
- through the waters (verse 2)
- in exchange for your life (verse 4)
- the waters... the rivers... the fire (verse 2)
- your Saviour (verse 3)
- summoned you by name (verse 1)
- precious (verse 4)
- Do not be afraid (verses 1,5)

Explain that these words were written for God's people held captive far from their own land. If God promised to be with them through that, we can be sure he'll be with us through something like starting at a new school.

Ask those changing schools to say where they are starting and how they think it might be different from their old school. Get some or all of the others to read out Isaiah 43:1-5 to them.

Schoolmates (10 minutes)

Make big wall posters as gifts to encourage group members who are starting at new schools. Use words or ideas from Isaiah 43:1-5. While some use craft materials, others might prefer to work on a computer.

Pray for these individuals. Simply have a time of silence after you say the name of each person – or have open prayer if the group are comfortable with that.

Idea sack

Design a school
☒ PREPARATION ☒ BOOK

Cover the floor with paper, e.g. using spare rolls of wallpaper. Assign two teams half the floor each in a competition to design the perfect school. They should use the whole area, drawing in the layouts of buildings and other features and labelling them. Set a time limit. Talk through the designs. Then get each person to stand in their favourite place in the imaginary schools and pray for the real schools they attend or are moving to.

To talk positively about school

School angels
☒ PREPARATION ☒ BOOK

Assign each person starting at a new school one or more 'angels'. These are group members who agree to pray for them each day in their first week and to phone up after the first day to see how it went. Some may be able to give more support, especially if they attend the same school, e.g. accompanying them on the first day.

To give practical support to those changing schools

Mapped out
☒ PREPARATION ☑ BOOK

If some group members already attend schools at which others are starting, ask them to draw a personalized plan of the school. An official school plan could be used as a basis. The group's version should include drawings or comments on different places – 'Where I broke my toe.' 'Good for footie.' Photocopy the maps for the new students.

To make a new school site more accessible

Express timetable
☑ PREPARATION ☒ BOOK

Ask three group members or leaders to prepare a dramatic sketch. One person is a speaking clock who says the hours through the day at fifteen second intervals, starting with '7 a.m.' The others mime at express speed what happens through a student's school day.

Ask those starting at a new school to say which parts of the day they are most and least looking forward to. Get other group members to set watch alarms to remind them to say instant prayers for them at those times.

To pray for those changing schools

Back beatitudes
☑ PREPARATION ☑ BOOK

Write out the beatitudes of Matthew 5:3-10 on cards. Put the first part of each verse on one side of the card. On the other write the connected promise reworded as an offer, e.g. 'Enjoy the kingdom of heaven!' (verse 3).

Place the cards 'offer' side up. Ask each pair of group members to choose one and turn it over. They must agree what the words mean (with help from you) and act out an example of how it could be put into practice at school.

To think how to live as Jesus' followers at school

Survivors' celebration
☑ PREPARATION ☒ BOOK

Three weeks after the start of term, hold a 'Survivors' celebration'. Get those who have changed schools to choose what they would like to do for this – this would be especially good if they are also new to your group. In the chatting at the event, find out how they are getting on.

To celebrate changing schools

Holidays

KEY BIBLE PASSAGE   *Matthew 5:13-16*

Why celebrate?

Holidays can be great opportunities to have a break, to chill out, to enjoy time with family or friends and experience new places and activities. The sabbath principle shows that time off is also part of God's pattern for us. But holidays can also be times of boredom, of temptation, of falling out with those close to us and falling away from God.

Jesus challenged his followers to be salt and light to the world – doing good so that God would be honoured. That is a full-time task – with no holidays. As well as celebrating the opportunities that holidays bring, we need to challenge our young people to use the time well, resist temptation, have good relationships and hang in there with God. And we need to make special efforts to pray for them through the break.

'You are the salt of the earth.' **Matthew 5:13**

Starter pack

Perfect holiday (10 minutes)

Beforehand, write up the following activities on an OHP acetate. Make an overlay sheet showing the scores. (Alternatively, use a flipchart with the scores hidden.)

	Score		Score
hang out with friends	2	shop till you drop	1
bike or hike	2	a big lie in	1
watch all the soaps	1	read trashy mags	1
a bumper prayer time	3	read the Bible	3
do nothing	1	chat with your family	2
nuke aliens on the computer	1	go to something at church	3
slave over school project	2	earn some cash	2
help with the hoovering	3	watch a decent video	2

Ask each group member to write down five activities to make up their ideal day in the holidays. Reveal the scores. Ask them to total the scores for their activities:

12–15	Holy, holy, holy!
6–11	Don't get bored!
0–5	Sad! And what about God?

Point out that the scores are just fun, but how we use our holidays as Christians does matter.

Salt and light (15 minutes)

Divide into groups of three to five. Read out Matthew 5:13. Ask the groups to suggest alternative foods which can replace 'salt' wherever it appears (e.g. chocolate, ketchup, ice). Get them to say their versions. Then read out verses 14-16. Explain that as salt has to stay salty, chocolate chocolatey and light bright, Christians need to keep being like Jesus. On holiday can actually be one of the hardest times to do it.

Give each group either Philippians 4:8 or Colossians 3:12-13 to look at. Ask them to imagine being on a family holiday or day out. How might they put the ideas in their verses into action? How might it all go horribly wrong? Give each group two postcards to write – one from the 'salty' holiday, the other full of the disasters. Read out the results.

Refreshing (10 minutes)

Invite individuals or pairs to mime the following refreshing holiday activities for the rest to guess. You could have teams and score.

- Diving into cold water
- Taking a shower
- Listening to cool music
- Ice skating
- Ice-cold can of drink
- Strawberry ice cream
- Bike ride in the rain
- Jumping waves in the sea
- Swimming under water

Point out that the holidays are also a good time to get refreshed with God. Put up the following list of holiday prayer opportunities:

- On the beach
- With a mate
- In the shower
- On a roller-coaster
- Wherever
- In a church
- On a bike
- Up a mountain
- In bed

Pray for refreshment during the holiday and for God's help to stick close to him.

Idea sack

Game glut
☑ PREPARATION ☒ BOOK

Have a holiday event where you bring a random collection of household items, games equipment and whatever junk you've got lying around at home. If possible have a supply of water. Split into groups of two to six and ask each individual in turn to choose an item from the pile. Then give them ten minutes to invent a game using all their items – and water if they want. Play all the games, however great or gruesome.

To build relationships and have a laugh

Holiday guide
☑ PREPARATION ☑ BOOK

Give all the group members copies of Proverbs 15 and highlighter pens. Allow them ten minutes to read them and to highlight the three which they think are most relevant to their coming break. Then, as a whole group, discuss and vote on the top three holiday guidelines from those highlighted.

To live wisely on holiday

Holy day
☑ PREPARATION ☒ BOOK

Put aside a whole day during a long holiday for the group to spend together. Run from breakfast to bedtime with a programme of activities: prayer, games, singing, swimming, food, a trip out, doing nothing, music, video, Bible reading (take the chance to tackle something bigger than usual), and so on. You could plan to invite others to join you for part of the day, e.g. invite friends along for a barbecue in the evening.

To be refreshed and challenged

Top hols
☑ PREPARATION ☒ BOOK

Ask the group members to flick through holiday brochures and decide their top five priorities for a dream holiday, e.g. watersports, good mates. Get them to make a 'dream holiday' postcard using brochure pictures or words, a plain postcard, scissors and glue.

Brainstorm ideas about spending time with God in Bible reading and prayer over the holidays: what to do, where and when. Ask each person to choose five top ideas and to write them on the back of their postcard to take away.

To pray and read the Bible in the holidays

In that case
☒ PREPARATION ☒ BOOK

At the last session before the long holiday, place an open suitcase in the middle of the floor. Fill it with items which relate to what you have done in the group in the last year, e.g. things on the walls of your meeting place. Use this to recap some of what you've done and learned. Close the suitcase to make the point that all those truths about God and about being a Christian go with you on holiday.

To recap and stay close to God in the holidays

Most boring postcard
☒ PREPARATION ☒ BOOK

Have a competition for the most boring postcard discovered over the holidays. To enter, members need to write them and send them to any other group member who must bring them along after the holidays. Get an independent judge to choose the winner. Include postcards of your local area as well as holiday destinations.

To keep contact over the holidays

Moving away

KEY BIBLE PASSAGE   Ruth 1:6-17

Why celebrate?

Moving house and starting again in a new community is one of the most exciting changes in life. But it is also one of the most stressful life-events we can face. As group members move on – to a new area or just out of the group – we can help them through the break, keep contact, and encourage them to get bedded into a new church or group.

For Ruth, moving from Moab to Judah meant much more than facing a new home. By going with Naomi she was taking on Naomi's faith – throwing in her lot with the true God of Israel. There was to be no turning back. We want our group members to cling to God with the same determination as Ruth – to rest under his protection wherever they go.

'Where you go I will go, and where you stay I will stay. Your people will be my people and your God my God.' **Ruth 1:16**

Starter pack

Tributes (15 minutes)

Use some of the following ideas to celebrate the group members who are leaving. Be sensitive if they are moving for a sad reason, e.g. family break-up – they'll need affirming, but don't trample on painful areas.

- Use a photo of them for a poster-size backdrop, a jigsaw or T-shirts.
- Play their choice of music.
- Interview them about their time in the group, what they've learned or seen God doing.
- Show photos (or videos) from the life of the group, with a caption competition for one.
- Display photos of the leavers at various ages.
- Ask others to relate incidents from the past: 'Remember when...'
- Talk about where they are moving to, e.g. show a photo of their new home.

- Give an 'Oscar'-type trophy affirming the specific qualities they've brought to the group.
- Do a mini 'This is your life'.
- Thank God for them.

Moving stories (15 minutes)

Read out this letter:

> *Dear...*
>
> *I'm 13. I go to a youth club in our village hall on Saturday nights. It's OK. Only eight or so people go, but they're my age. The club's just got some lottery money so they'll be getting some new equipment. It isn't Christian or anything, mind you.*
>
> *My best mate, Nathan, is 15. He's decided to start going to the 'Zone 1' events in town on Saturday nights – ten miles away. It's a church thing for 13–18s with talks and decent bands. But no one else under 15 goes.*
>
> *Should I go with Nathan or stick where I am?*
>
> *Russ*

Ask the group members what Russ should do and why. Get the arguments on both sides.

Explain the background to Ruth's story (as in Ruth 1:1-5). Read Ruth 1:6-17. Ask: What was most important to Ruth in her decision? Make the point that Ruth was faithful to Naomi. She was willing to live anywhere – but was determined to stick with the true God. We need to be determined to stick with God whatever moves and changes we go through.

Past to future (10 minutes)

Give each group member a card shaped like a large bookmark. At the top write 'Birth' and at the base write 'Now'. Ask the members to list on it people, places and situations they have left in the past, e.g. homes, friends, schools.

Give them the following Bible references to look up: Psalm 136:26; Matthew 28:20; Jeremiah 29:11; Philippians 2:6. Invite them to choose one and write it out on the reverse of the bookmark as a reminder that whatever changes they face, the future lies with God.

Idea sack

Going, going, gone!

☑ PREPARATION ☒ BOOK

Use this where members are moving up to another age-group. Sit in a circle, with a wooden mallet on a table in the middle. Ask each person in turn to speak for fifteen seconds saying what they think is special about the group. Those who are moving up should say what they will miss. After fifteen seconds shout 'Going, going, gone!' and bash the mallet on the table. The next person must start immediately.

To celebrate the life of the group

Shelter and strength

☒ PREPARATION ☑ BOOK

Split the group into threes and ask them to look at Psalm 46:1-3. It talks about God being our shelter in times of trouble. Assign each group one of these situations:
- a small child starting at school
- an 18-year-old leaving home
- a family moving to be mission partners in a desert area
- an astronaut before take-off
- (*group member*) moving to (*place*).

Ask them to rewrite the verses for those people using words and images which relate to their situations. Read them out.

To strengthen trust in God when moving away

Bless 'em

☒ PREPARATION ☑ BOOK

Get the group members to memorize the words of the blessing in Numbers 6:24-26. Say them together to those leaving at the end of a 'farewell' social event or to finish their last regular session with you.

To send off those leaving with the assurance of God's care

Move out

☑ PREPARATION ☒ BOOK

Have a session where you start in your usual meeting place then move to a venue the group members have never been to. Take a few familiar items with you. Use this to get the group members to reflect on the process of moving. Ask what they like about a change of venue and what they don't like. Have the rest of your session there.

To reflect on the experience of moving

Memory box

☑ PREPARATION ☒ BOOK

Prepare a small box or album of items which will help the person moving away to remember the group. Include photographs, items related to activities you have enjoyed together or the life of the group, signatures or messages from group members, favourite Bible verses or prayers. Make planning and putting this together a group project.

To help someone leaving remember the group

Write back here

☒ PREPARATION ☑ BOOK

Get the group members to devise a fun questionnaire for the person moving away to fill in and send back to you about a month on. They might ask about their new home, school, church, friends, feelings, etc. Get them to write it with a choice of answers for each question, e.g. 'What is your new area like?' a) Neighbours b) Emmerdale Farm c) Coronation Street d) other…

When it is returned, pray together for the person concerned and write back.

To keep links and pray for someone who's moved

New group members

KEY BIBLE PASSAGE | *Ephesians 4:1-6*

Why celebrate?

It is normal to feel self-doubt and nerves when arriving as a newcomer at an established group. Celebrating the arrival of new members in the group will show them that they are really welcome – that they are not looked down on but are valued from the start. Where several people come up from a younger age group, you can plan a suitable session for the first week. Individuals trying out the group at other times need to be equally welcome but also allowed space to be 'just looking'.

Paul reminds the Ephesian Christians of the amazing things which unite them – the Holy Spirit, faith in Jesus, the sovereign God working in them all. The members of our groups have the same brilliant things in common. That should lead to peace and to humble, patient, loving relationships. There's lots to celebrate and work on!

'There is… one God and Father of all, who is over all and through all and in all.'
Ephesians 4:4,6

Starter pack

In common (10 minutes)

Include leaders in this. Get into pairs. The two people must try to find likes and dislikes which they have in common. Get them to count how many they can find in one minute. Get into threes or fours. Again they must try to find likes and dislikes which all of them have in common. Repeat, increasing the size of the groups as appropriate. For the last go, work together as a whole group, writing up likes or dislikes you all share on two lists: 'We like…' and 'We don't like…'

Ask: What do all Christians have in common? Listen to their ideas then read out Ephesians 4:1-6 in a simple, direct translation, e.g. *The Message* or *Contemporary English Version*.

One (15 minutes)

Read out Ephesians 4:1-6 making some deliberate mistakes: e.g. 'poisoner' for 'prisoner', 'proud' for 'humble'. The group members must spot the errors and give the correct version – perhaps working in teams and scoring.

Point out that all Christians have such amazing things in common. Discuss: Does that make them behave differently towards each other? If the group members say 'yes', encourage them to give specific examples from their experience.

Ask the group members to suggest and agree ground rules for your group, e.g. 'We will listen carefully when someone else is speaking', 'We will pray for each other between meetings.' Encourage them to include ideas from the passage. Write up those ground rules which they all agree to follow and stick them on the wall. If you have a written aim for the group already agreed by the group leaders or church leadership (and you should!), include this on the sheet.

Welcome ideas (10 minutes)

Provide slips of paper of two different colours. Invite the group members to write on one colour of paper how they felt when they first joined the group. (For newcomers that will be now.) On the other they should jot down ideas of how the group could help the new members settle in. The newcomers should give their suggestions, too.

Sit in a circle. Redistribute the comments at random so they are anonymous – one of each colour for each person. Throw a ball round the group. Whoever gets it must read out their 'feelings' slip. Repeat the ball throwing, reading out the practical ideas. Plan how you can put some of these into action – perhaps at a special event.

Pray for each new group member by name, and for the whole group using ideas from Ephesians 4:1-6.

 Idea sack

Exploding name game

☒ PREPARATION ☒ BOOK

Sit in a circle. Start by saying 'My name is… and I enjoy…' The next person must repeat your information and then give their own: 'His/her name is… and he/she enjoys… My name is… and I enjoy… *and*…' The third person repeats all the previous information and says *three* things they enjoy. When it crashes, start again changing the line to 'I *don't* enjoy…', 'I know…', 'I have…', 'I can…', 'I can't…'

To get to know each other

Posh nosh

☑ PREPARATION ☒ BOOK

A few weeks after an influx of new members, hold a posh sit-down dinner for the group. Do formal invitations. Get others from the church involved as cooks and appropriately dressed waiters. You could have a menu with two or three choices and get the members to choose beforehand. Provide non-alcoholic cocktails. Let group members plan the background music.

To get to know each other and feel special

Body bits

☑ PREPARATION ☒ BOOK

Play hangman without drawing gallows – instead put on body parts, e.g. plastic ears, false nose, spectacles with eyeballs, wig, moustache, beard, gloves and flippers – eleven items equivalent to a complete gallows. Let group members take turns. The phrases to guess must be groups of people, e.g. 'The All Blacks'.

Write out 1 Corinthians 12:12,15,18,21,26. Look at each verse and ask what that image of God's people might mean for you as a group.

To relate to each other as God's people

Hot hi!

☒ PREPARATION ☒ BOOK

When new members are due to join the group, get the others together the week before in someone's home to plan a really warm welcome. Brainstorm what makes you feel welcome when you visit someone's home. When you've written up lots of ideas, think how some can be adapted for your group situation. Then get everyone involved in putting it into action in the next week.

To welcome new members

Bursting with questions

☑ PREPARATION ☒ BOOK

Think up lots of questions for group members to answer about themselves, e.g. 'What's your favourite place?' 'What would you do with $/£1,000?' 'What was your most embarrassing moment?' Put each one in a balloon and blow it up.

Get someone to burst a balloon and ask the question inside to anyone of their choice. After answering, it is that person's turn to burst a balloon. Each person is allowed one 'cop out' – when everyone else in the group must answer instead.

To get to know each other

Christ be with us

☑ PREPARATION ☒ BOOK

Explain that Jesus is the reason your group exists. He cares desperately about it and will be at work in it. Say together the following, based on words of St Patrick:

'Christ be with us, Christ within us; Christ behind us, Christ before us;
Christ beside us, Christ to win us; Christ to comfort and restore us;
Christ beneath us, Christ above us; Christ in quiet, Christ in danger;
Christ in hearts of all that love us; Christ in mouth of friend and stranger.'

To focus on Jesus

Party bag

Party planner

✗ PREPARATION **✓** BOOK **✓** ANY OCCASION

Spread round the room six sheets of paper with the following headings: food, people, times, places, activities, reasons to celebrate. Give the group members five minutes to write on the sheets their ideas for a perfect party.

Read out the 'reasons to celebrate', getting the group members to vote or shout for those they like. Run through the other sheets deciding which ideas would fit best with the chosen 'reason'. Then plan the party.

To build relationships

Letter cake

✓ PREPARATION **✗** BOOK **✓** ANY OCCASION

Hold a special session in someone's home when the group members make cakes of different kinds in square cake tins. Get them to cut the cakes and use the pieces to spell out a message from them to the rest of the church appropriate to the occasion.

To celebrate with the rest of the church

Celebrate with us

✓ PREPARATION **✗** BOOK **✓** ANY OCCASION

As a first contact point, celebrate something that matters to group members' friends, e.g. success of a local sports team (or failure or mediocrity), a pop-group visit or break-up, a big event in a soap. Get the whole group behind the event, even if it's not their thing. Plan other celebrations to appeal to different friends. Don't use it for up-front evangelism; instead, invite the guests to another group meeting or event.

To help young people make contact with the group

Times challenge

✗ PREPARATION **✓** BOOK **✓** ANY OCCASION

Get hold of a range of hymn books from your own and other churches. Split the group into smaller ones, and give them ten minutes to find as many references as possible to seasons and time passing. Once the time is up, allow the groups in turn to read out the relevant verse of any hymns they have found. Have a go at singing one or two of them if possible.

To help the group understand the heritage of worship

Time link

✓ PREPARATION **✗** BOOK **✓** GROUP BIRTHDAY

For a special celebration of, say, ten, twenty-one or fifty years of your group's existence, track down as many as possible of the original group members and invite them to come to a group session. Consult with them beforehand about what used to happen at the group when it started. Plan the session to start and end with your typical activities but make the middle part like those original group times.

To see God's work in the group past and present

Celebrate by numbers

✓ PREPARATION **✗** BOOK **✓** GROUP BIRTHDAY

For a tenth or twenty-first birthday, plan to have ten or twenty-one celebratory events through the year. Each one should be different.

To build relationships and celebrate God's work

Musical choice

☑ PREPARATION ☑ BOOK ☑ GROUP BIRTHDAY

Arrange for the group members to choose the songs (or at least some of them) for the service of your 'official' group birthday. At least two weeks beforehand, invite the service or worship leader along to say how they choose songs, what books can be used, and so on. Get the group members to work on brief explanations of why they have chosen them. Use this in the service to introduce the songs.

To communicate the group members' faith to the wider church

Rev – elation

☑ PREPARATION ☒ BOOK ☑ ARRIVAL OF NEW MINISTER

Invite the new minister to visit the group in their first few weeks. Interview them about:
- themselves when they were the age of the group (their home, church links, etc.)
- how they came to be your minister
- how they see their job in the church.

Pray for them. Give them a box of Quality Street but with some wrappers containing welcome messages written by group members instead of chocolates.

To welcome a new minister

Rave review

☑ PREPARATION ☒ BOOK ☑ HALF-WAY THROUGH THE YEAR

At the mid-point of the group's year combine a 'not half' celebration with a programme review. For the celebration have halves of food items and music tracks, each dress half-and-half in two styles, think about God's whole-hearted love (in some places this will coincide with Valentine's Day). For the review get the group members to comment openly on the programme so far and give their creative ideas. Address their suggestions wisely and adventurously.

To celebrate the group's life and possibly Valentine's Day

Summer shade

☒ PREPARATION ☑ BOOK ☑ SUMMER

This is like 'Winter warmer' for summer and hot climates. You need an 'overheat' location and somewhere shady and cool. In the first, read Psalm 42 and pray for people who are oppressed in any way. Then in the shady place read Psalm 121 and pray for God's protection for Christians in danger.

To pray for people suffering in summer

Winter warmer

☒ PREPARATION ☑ BOOK ☑ WINTER

Set up a summer room which is hot and bright with sun hats, sun cream and so on in it, and a winter room which is cold and dark with white sheets covering anything bright. In the summer room read Job 37:14-23 – words for someone who had suffered terrible things. Pray for people who have suffered personal disasters. Then go to the winter room, read Lamentations 3:1-20 and pray for those who are sad or depressed at this time.

To pray for people suffering in winter

Dummy visit

☑ PREPARATION ☒ BOOK ☑ VISITOR TO CHURCH

When a distinguished visitor is coming to the church (e.g. a bishop), ask if they can visit the group to see a typical session. The previous week do a dummy run of some activities you will use – taking turns to play the role of the visitor (denoted by a dog-collar or mitre). Revise plans to make sure they are really included and welcomed, including praying for them.

To welcome a visitor and show them group activities

Squirt circle

❌ PREPARATION ❌ BOOK ✅ NEW GROUP MEMBERS

Stand in a circle, well apart, each person facing the back of the next person. Pass a can of 'silly string' round the group. Each person must squirt the one in front saying '(*name*) this is a squirt of (*any substance*)'.

To build relationships

Age-group dip

✅ PREPARATION ❌ BOOK ✅ NEW GROUP MEMBERS

When new group members are moving up from an age-group below, invite them on a trip to a swimming pool in the week before your first session. If possible, go for a leisure pool with flumes and waves. Depending on numbers you could pay for them. Afterwards return to base for some food.

To help group members transfer up

Beano annual

✅ PREPARATION ❌ BOOK ✅ CHURCH AGM

Brighten up the church annual general meeting by asking if the group can report on their year as a rap or with drama or music. They should include what they have done, achievements, struggles and issues for the wider church. Get them to write a clear straight report too. They might start:
'Listen up for some news, brothers. Sisters, listen well.
To the things we bin doing and the message that we tell.'

To inform the wider church about the group

Ecumenicrawl

✅ PREPARATION ❌ BOOK ✅ ANNIVERSARY OF DENOMINATIONAL CONFLICT

Celebrate an anniversary of denominational conflict, e.g. 5 November in the UK, by having an ecumenical 'grub crawl'. The group members from one or more churches go from place to place for each course of a meal. Either use homes of people from the different denominations, or the church buildings themselves. Include some activities which will help people to get to know each other.

To build relationships with other churches

Country view

✅ PREPARATION ❌ BOOK ✅ A NATIONAL DAY

Bring along magazine pictures of your country, showing contrasting aspects of life: people, buildings, church, scenery, government. Spread pictures out in the middle of each group of three to four. Ask them to pick out things they like in the pictures. All move to another set of pictures and discuss what they dislike. Move again then pray three-word prayers using these pictures to spark off ideas. Alternatively use just one set of pictures.

To pray for your nation

Daily diary

✅ PREPARATION ❌ BOOK ✅ CHURCH OR GROUP HOLIDAY

Put together a sheet with spaces for the members to write in labelled 'Star spot', 'The pits', 'I learned…', 'Someone said…', 'Favourite place', 'I met…', 'God…' The spaces could be appropriate shapes (star, dustbin, etc.). At the end of each day of the holiday, get the group members together. Ask them to fill in the sheet with words or drawings about their day and then to tell each other how they are getting on.

To support each other on a group holiday

Who are we?

❌ PREPARATION ❌ BOOK ☑ NEW TERM

At the start of a term look at the name of the group and ask the members if it reflects who you are and why you exist. If not, encourage them to come up with a better name and a logo. Help them to consider what messages the proposed name might communicate to their friends and others in the church.

To strengthen group identity

Shout to God

❌ PREPARATION ☑ BOOK ☑ TESTS AND EXAMS

Ask the group members to imagine that they are on the test run of a new deep-sea diving craft. They get down to an incredible depth. It's very dark. But then they hear a dripping sound. The craft is leaking. Ask them to write down the message they would yell up the communication line. Put the messages on the floor and together arrange them into a prayer to use in tough situations.

To create words to pray in tough situations

Water retreat

☑ PREPARATION ☑ BOOK ☑ AFTER EXAMS

Organize a day away as a retreat after exams or during the summer break. Go to a peaceful place by a lake, a river or the sea, if possible. Have a time for reflection, look at some Bible material such as Psalm 93 (God's power and the sea), the river of life in Genesis 2, Ezekiel 47 and Revelation 22 or 'learning about Jesus round Galilee' in Mark 1-8. Then go wild with beach or water-based games.

To unwind and learn about God

Weekenders

☑ PREPARATION ❌ BOOK ☑ AFTER EXAMS

Use whatever contacts you (or your minister) have to find a church with a similar size of group to yours in another part of the country. Invite them to come and visit you for a weekend after exams are over. Either do a joint sleepover or fix for them to stay with members of the group. Plan a packed programme for Saturday with a relaxed barbecue in the evening at someone's house. Keep in touch and arrange a return visit.

To build relationships with Christians elsewhere

Imperfect parents

☑ PREPARATION ❌ BOOK ☑ PARENT CELEBRATIONS

Record soap opera clips showing relationships between parents and children. Ask the group members to say when they spot where those relationships are working well or badly. Hand a bar of soap round, each person completing the sentence 'When I'm a perfect parent, I'll…' Repeat with 'But I won't be a perfect parent because…' Pass the soap round a last time saying one thing they would like to pray for all parents.

To understand and pray for parents

Dramatic prayer

☑ PREPARATION ❌ BOOK ☑ ANY CHURCH FESTIVAL

Choose a prayer which fits a special occasion when the group members will be in the main church service (e.g. the appropriate Anglican collect). Write out each phrase of the prayer. Discuss with the group what it means, and maybe get them to rewrite it in their own words. Together devise some dramatic action or a visual image to go with each phrase. Work on these and use them in the service.

To help others pray

Birthday sellaway

☒ PREPARATION ☒ BOOK ☑ CHURCH BIRTHDAY OR DEDICATION FESTIVAL

Swap stories about birthday presents you have not liked or needed. Encourage the group members (with parents' permission) to bring along any unwanted presents from past birthdays. Have a 'Birthday Sellaway' stall selling them to church members, friends or each other. Give the money to the church for a specific use, e.g. toys for a toddler group, a mission partner or building project.

To give to a church project

Bursting for joy!

☑ PREPARATION ☒ BOOK ☑ HARVEST

Provide lots of balloons. Talk about good things in life – friends, food, health, the group, etc. Give them five minutes to write as many specific examples as possible on the balloons with permanent markers or OHP pens.
　　Read out one item from each balloon. Each time the group should respond: 'Yes, thank you, Lord'. Finish by letting them burst the lot (or hang them up round the room – the balloons, that is).

To thank God for good things

Crunch questions

☑ PREPARATION ☒ BOOK ☑ HARVEST

Have a church quiz night. Some group members might help organize it. Have mixed teams of younger and older people and questions which draw on the knowledge of different age-groups. Possible rounds would be: identifying music clips, TV and films, facts about your church and building, identifying recorded sounds (Jim eating breakfast), Bible characters, the local area (what road is ... on), natural world, sport. Set the task of writing down as many countries as possible through the evening.

To build relationships in the church

Luke check-up

☑ PREPARATION ☒ BOOK ☑ ST LUKE'S DAY (18 OCTOBER)

Give two weeks' notice that you will celebrate St Luke's day with a team quiz on Luke 5. Encourage the group members to read it in advance. The questioner should dress as a doctor (i.e. Luke) and say things like 'Now you've come to see me for a little examination.' Mix the questions with physical 'medical' tests, e.g. transferring dried peas with a straw.

To enjoy Luke's Gospel

Has beans

☑ PREPARATION ☑ BOOK ☑ ALL SAINTS

Write on cards all your church's groups and activities, e.g. drama group, morning services, church spring cleaning, church prayer meeting. Put a bowl by each. Give each group member twenty kidney beans to distribute as they want, to indicate how relevant they think the activities are to them.
　　Look at the results together. Discuss why they scored them as they did. Are there any activities which they would like to be more involved in? Plan what action to take.

To get involved in church life

Changed names

☒ PREPARATION ☑ BOOK ☑ ALL SAINTS

Ask group members to list the names of various members of the church, of whatever age. They should also collect some they don't know, especially newer people. Get those who like this sort of thing to form anagrams of the names. (Someone may have some anagram software you could use.) Make the best ones into a quiz sheet for a church social event or just for fun after a service.

To get to know other church members

Character meal

☑ PREPARATION ☑ BOOK ☑ BIBLE SUNDAY

Beforehand, give each person an invitation showing a Bible character role, two sentences about them, costume suggestions and a couple of Bible passages for more information.

First explain the scenario: 'You're outside heaven's gates, shut out due to a computer error. You must each say why you should get in.' Have three rounds alternating with courses of a meal. Vote after each round (two votes each and no one may vote for themselves). The first and second times half the candidates are eliminated; the third vote gives the winner.

To get to know Bible people

Bible time line

☒ PREPARATION ☑ BOOK ☑ BIBLE SUNDAY

Ask the group members to write or draw on 'Post-it notes' their favourite Bible bits – people, events, stories, verses. Stick these up on a wall. Then work together to rearrange them into time order. Get the group members to do as much as possible using helps in their Bibles. When you've finished, challenge them to try telling the whole story of the Bible in less than one minute.

To get an overview of the Bible

Eve lights

☑ PREPARATION ☒ BOOK ☑ CHRISTMAS EVE

Invite any group members who are around on Christmas Eve to evening meals in leaders' homes. The aim is to have a 'family' feel, so only have a few members in any home, and include other household members or invite one or two others from church. Keep it simple. After eating, sit by candlelight and read Matthew 1:18-25 and Luke 2:1-20.

To sense the wonder of Christmas together

Claus encounters

☑ PREPARATION ☑ BOOK ☑ CHRISTMAS

Write out the following sentences on cards. The group members must make silly sentences about Santa Claus, then read John 3:16, Ephesians 1:6-7 and Romans 8:32 for the truth. (This may seem an irreverent exercise, but the point it makes can be very effective.)
1. 'For … so loved … that he gave …, that … shall …'
2. 'his glorious …, which he has … given us in … . In him we have …'
3. 'He who did not … but gave … how will he not also … give us …?'

To contrast a materialist view of Christmas with God's great gift

Hark the harold

☒ PREPARATION ☑ BOOK ☑ CHRISTMAS

Give out photocopies of the first verse of 'Hark! the herald angels sing'. Ask the group members to circle any words they think could be made simpler. Compare notes. Share the lines out for pairs to rewrite in words they might use – they don't have to scan or rhyme but they should keep the meaning. Put it all together. One or two people could try to mould it into something which does scan and put some new music to it.

To use and understand Christmas songs

Problem page

☑ PREPARATION ☑ BOOK ☑ LENT

Invite the group members to write down questions and problems that they or others in their peer group face concerning life as a young person and as a Christian. Explain that two named people (perhaps including you) will write replies, but no one else will see them. Select someone wise with knowledge of the age-group to write sensitive and brief replies – and another to check what they've written.

To provide a forum for young people to ask questions.

Shouts and tears

✗ PREPARATION ✗ BOOK ✓ PALM SUNDAY

Read Luke 19:28-44. Give each group member a 'role': Jesus, two disciples, colt's owners, crowd, Pharisees. As you read it again, ask them to close their eyes and imagine the scene through the eyes of their character. At the end ask each person to speak one sentence to express what their character is feeling.

Praise Jesus as King in a song or shout. Then pray silently for people who do not recognize him as King.

To worship Jesus wholeheartedly as King

Shop till you drop!

✓ PREPARATION ✗ BOOK ✓ EASTER

Save money over a few weeks through donations, sponsored activities and excess dinner money. Split it equally between all the young people and invite them to spend exactly that amount on gifts for some of the older church members. Take advice from those who visit those older people about the kind of thing different individuals might like. Organize a shopping trip. Wrap and give the Easter presents.

To learn about giving and encourage older church members

Joy!

✓ PREPARATION ✗ BOOK ✓ EASTER

Get the group members to take lots of photographs of themselves in very different poses – half expressing joy, the rest showing other possible reactions to Jesus' resurrection – wonder, shock, doubt… Include close-ups, groups and long distant shots. Together plan a sequence of the pictures with appropriate music (see CD) to use in church at Easter. Transfer them to OHP slides by photocopying or scanning into a computer. (Or use transparency film initially.)

To respond to Jesus' resurrection

Praise all nations

✗ PREPARATION ✗ BOOK ✓ MISSION FOCUS WEEKEND/PENTECOST

Repeat Psalm 117 together until the group members can remember it. Ask them to think of some nations they are saying it to and how they themselves have seen God's love and faithfulness recently. Use it in church with the group, saying the first half of each verse and the congregation the second half. Repeat it over and over. If there are enough group members, start together in the centre, then spread out round the church and stop on an agreed signal.

To praise God for his love

Trinity party

✓ PREPARATION ✗ BOOK ✓ TRINITY SUNDAY

Hold a 'threes' themed party. Have music by bands with three members, food arranged in sets of three, play three games, invite three guests, wear three socks, shoes, etc. At the end point out that although the idea of the Trinity is about three persons, it is also about one God. Finish by standing in one circle for a closing prayer, thanking God for all the ways he reveals himself to us.

To celebrate God's nature

Three-in-one snaps

✗ PREPARATION ✗ BOOK ✓ TRINITY SUNDAY

Get three groups to pose themselves in snapshots of: Exodus 14:21-22, Mark 4:39 and Acts 4:8, looking at the verses around to understand what was happening. These verses show God the Father, Jesus and the Holy Spirit. Then read John 14:10, Matthew 4:1 and Matthew 10:19-20. These show: the relationships – Father and Son, Son and Spirit, Father and Spirit. Finally do a snapshot together of Matthew 3:16-17, deciding how to represent God the Father and the Holy Spirit with Jesus.

To explore the Trinity

Index

Party bag activities for:

Resources

General resources for leaders

CPAS code

C18001Y	Groups Without Frontiers	Terry Clutterham, Penny Frank, Phil Moon
03648Y	The Adventure Begins	Terry Clutterham (CPAS/Scripture Union)
C18005	The ART of 11-14s	Terry Clutterham, Jo Horn

Worship resources

C20010Y	DIY Worship (including CD ROM)	Simon Heathfield
C93400Y	Telling Tales	Dave Hopwood
C93403Y	Telling More Tales	Dave Hopwood
C93402Y	See What I Mean	Jonathan Mortimer
C16144Y	Drama Verses Sketches	Steve Tilley/Bob Clucas

Teaching resources

C25124Y	Absolutely Everything!	The whole Bible in eleven sessions
C16131Y	You'd Better Believe It!	Christian doctrine
C16142Y	You'd Better Believe This Too!	More Christian doctrine
C16141Y	Another Brick in the Wall	Nehemiah
C16127Y	All Together Forever	Ephesians
C16129Y	Pressure Points	Issues that concern teenagers
C16132Y	Repeat Prescription	The ten commandments
C16134Y	Powered Up	Key moments from Acts
C16135Y	Just About Coping	Issues that teenagers have to cope with
C16136Y	Mission in Action	Broadening your group's horizons
C16137Y	People with a Purpose	Ten Old Testament characters
C16139Y	Outlawed by Grace	Galatians
C16138Y	Didn't He Used to be Dead?	Jesus
C16146Y	The End is in Sight	Revelation
C16147Y	Me You Us Them	Relationships
C25123Y	Loose the Juice!	Parables in Luke's Gospel
C26001Y	PrayerZone	Ways to energize your group to pray
C16148Y	Where Do We Begin?	Creation and beginnings

More books from CPAS to resource your sessions

C16140Y	Know Ideas! 2	More ideas
C16143Y	Know Ideas! 3	And still they come!

All of these resources are available from
CPAS Sales, Athena Drive, Tachbrook Park, Warwick CV34 6NG
Tel/24-hour voicemail: **(01926) 458400**
CPAS website: www.cpas.org.uk